Routledge Revivals

Maritime Boundaries and Ocean Resources

First published in 1987, *Maritime Boundaries and Ocean Resources* is a collection of essays which examines the political jurisdiction of ocean boundaries and the affects that this has on the world's oceans. It examines how the intensification of ocean use has raised questions of how rational planning, and the management of the oceans can avoid increasingly environmental damage and sea use conflict and examines the ocean as a tool for space, trade and communication. It also addresses the creation of integrated regional planning for ocean management.

T0134112

Maritime Boundaries and Ocean Resources

Edited by Gerald Blake

Routledge
Taylor & Francis Group

First published in 1987
by Croom Helm and Barnes and Noble Books

This edition first published in 2018 by Routledge
2 Park Square, Milton Park, Abingdon, Oxon, OX14 4RN
and by Routledge
711 Third Avenue, New York, NY 10017

Routledge is an imprint of the Taylor & Francis Group, an informa business

Publisher's Note
The publisher has gone to great lengths to ensure the quality of this reprint but points out that some imperfections in the original copies may be apparent.

Disclaimer
The publisher has made every effort to trace copyright holders and welcomes correspondence from those they have been unable to contact.

A Library of Congress record exists under LCCN: 87000907

ISBN 13: 978-0-8153-5374-4 (hbk)
ISBN 13: 978-0-8153-5376-8 (pbk)
ISBN 13: 978-1-351-13551-1 (ebk)

Maritime Boundaries and Ocean Resources

Edited by
GERALD BLAKE

International Geographical Union Study Group on
The World Political Map

BARNES & NOBLE BOOKS
Totowa, New Jersey

© 1987 Gerald Blake
First published in the USA 1987 by
Barnes & Noble Books,
81 Adams Drive,
Totowa, New Jersey, 07512
ISBN 0-389-20726-8

Library of Congress Cataloging-in-Publication Data
Maritime boundaries and ocean resources.

Bibliography: p.
Includes index.
1. Economic zones (Maritime law) 2. Territorial
waters. 3. Marine resources conservation — Law and
legislation. I. Blake, Gerald Henry. II. International
Geographical Union. Study Group on the World Political
Map.
JX4131.M295   1987   341.4'48   87-907
ISBN 0-389-20726-8

CONTENTS

Acknowledgements

Foreword by R.J. Johnston

Preface

Notes about the authors

Chapter 1   Worldwide maritime boundary
delimitation: the state of play   1
GERALD BLAKE

Chapter 2   Delimitation of maritime
boundaries: emergent legal
principles and problems........   15
PATRICIA BIRNIE

Chapter 3   Straight and archipelagic
baselines.....................   38
J.R.V. PRESCOTT

Chapter 4   The importance of geographical
scale in considering offshore
boundary problems.............   52
EWAN ANDERSON

Chapter 5   The limits of the area beyond
national jurisdiction - some
problems with particular reference
to the role of the commission on
the limits of the continental
shelf   63
PIERS R.R. GARDINER

Contents

Chapter 6   Maritime boundaries and the
            emerging regional bases of world
            ocean management................   77
            HANCE D. SMITH

Chapter 7   Common fishery resources and
            maritime boundaries: the case
            of the Channel Islands   89
            STEPHEN R. LANGFORD

Chapter 8   European, national and regional
            concepts of fishing limits in the
            European Community............. 117
            MARK WISE

Chapter 9   Offshore jurisdictional claims of
            the Republic of Ireland........ 133
            PROINNSIAS BREATHNACH

Chapter 10 Maritime boundary problems in
            the Barents Sea................ 147
            R.R. CHURCHILL

Chapter 11 The United States Exclusive
            Economic Zone: mineral resources 162
            FILLMORE C.F. EARNEY

Chapter 12 Historical geography and the
            Canada-United States seaward
            boundary on Georges Bank........ 182
            LOUIS DE VORSEY

Chapter 13 Maritime boundaries in the
            Mediterranean: aspects of
            cooperation and dispute......... 208
            NURIT KLIOT

Chapter 14 Defining the indefinable:
            Antarctic maritime boundaries   227
            GERARD J. MANGONE

Chapter 15 Beyond the bounds? A consideration
            of local government limits in the
            coastal zone of England and Wales 244
            JOYCE E. HALLIDAY

Select Bibliography   257

Index   272

ACKNOWLEDGEMENTS

Thanks are due to our four Chairmen at Reading – Jim Bird, Ron Johnston, Bob Smith and Ewan Anderson – who kept the programme on schedule and encouraged lively debate. The collaboration of many others in the preparation of this volume is also gratefully acknowledged, especially Stephen and Olivia Langford who prepared the bibliography and index speedily and carefully, and Topline Typists of Durham who prepared the camera-ready copy with great skill. Several authors have also made personal acknowledgements in the notes at the end of their chapters.

G.H.B.

FOREWORD

R.J. Johnston

Political geography is enjoying a welcome
resurgence, as geographers of all subdisciplinary
persuasions realise the need for a full appreciation
of the nature and role of the state in human
affairs. To provide a focus for that work, steps
were taken in the early 1980s, led by John House, to
promote the establishment of a Commission on
Political Geography within the International
Geographical Union. A major international conference
was convened in Oxford by John House in 1983, just
six months prior to his untimely death, at which a
proposal for a Commission was written. The focus of
its work was to be an understanding of the World
Political Map, and I had the honour of election as
Chairman-designate.
    At about the same time, the Executive Committee
of the International Geographical Union initiated a
new organisational category of Working Groups, which
were to have a limited life of four years during
which the case for establishment of a full
Commission could be made. Consequently the case was
made for establishment of a Study Group on the World
Political Map, and this was agreed at the Paris
International Geographical Congress in 1984.
    During the four-year life of the Study Group a
full programme of meetings and conferences has been
arranged in several parts of the world, each
focusing on a particular topic. It is hoped that a
number of publications will stem from these
meetings, helping to inform the geographical
community at large of our work and the contributions
that it is making.
    The first book to be published from the Group's
meetings was a selection of papers from the 1983
Oxford conference. (P.J. Taylor and J.W. House,
editors, <u>Political Geography: Recent Advances and</u>

Foreword

<u>Future Directions.</u> Croom Helm, London, 1984). The present volume is the second, and results from a meeting held in conjunction with the annual conference of the Institute of British Geographers at Reading in January 1986. Its topic is not only a crucial one in contemporary world affairs but also of great importance to political geographers since it involves a re-evaluation of the concept of territory as it applies to the state. Today, the earth's oceans are being claimed as part of the territory of sovereign states, along with the land surface. The chapters in this book illustrate the many issues involved and the interdisciplinary work that is being undertaken.

On behalf of the Study Group, I thank Gerald Blake for all that he has done to bring this book to fruition. It is the first of a series that we hope will emanate from our work, and will lead to a clear recognition of the importance of the perspective provided by political geographers.

Sheffield

PREFACE

Gerald H. Blake

This volume arose out of a meeting on maritime
boundaries and ocean resources initiated by the
International Geographical Union's Study Group on
the World Political Map. The Institute of British
Geographers kindly allowed the meeting to coincide
with their annual conference at Reading in January
1986, thus facilitating attendance by a good number
of geographers. Contributions to the cost of the
meeting were made by British Petroleum, Shell
International, and the Government and Law Committee
of The Economic and Social Research Council, whose
collective generosity is gratefully acknowledged.
Papers presented at Reading have been supplemented
in the following pages by five contributions from
other specialists in the field of maritime
boundaries and resources.
    The meeting proved to be an extremely useful
encounter chiefly between lawyers, geographers, and
hydrographers, who were agreed that no single
approach can hope to solve all the complex problems
associated with maritime boundary delimitation. The
task of drawing the offshore political map of the
world, and evaluation of its implications for the
allocation and management of resources will occupy
experts in many disciplines for decades to come.
Discussion at Reading led to the tentative
identification of six broad areas in which
professional geographers might make valuable contributions. Both the content of the following pages,
and several of the bibliographical references cited,
indicate that a vigorous start has already been
made, but much remains to be done. Beginning with
the two most important, the six areas of potential
geographical activity are:
    (a) Objective definitions of geographical
terms. The 1982 UN Convention on the Law of the Sea

adopts a number of geographical terms which are never satisfactorily defined. Examples are 'natural entrance points' of bays, 'rocks' and 'islands', and 'general direction' of the coast. Much could be done to identify the problems in using such terms, and in attempting objective definitions.

(b) Objective measurement of geographical characteristics. Geographical techniques could be divised to measure characteristics such as the degree of oppositeness of coastlines, coastal orientations, and the effect of scale on such measurements.

(c) Cartographic and surveying problems. Many technical problems are encountered when using maps and charts for maritime boundary delimitation. L. Alexander, P. Beazley and R.W. Smith have written about the implications of different choices of scale, projection, straight lines, spheroids and horizontal datums. Hydrographers, geographers, and surveyors might usefully demonstrate the effects of such choices in selected cases.

(d) Regional studies of boundary delimitations. The process of martime boundary delimitation is powerfully conditioned by regional considerations (physical characteristics, inter-state relations, the use of ocean space, seabed resources etc). Geographical studies of regional seas highlighting distinctive local factors might encourage regional solutions to boundary problems.

(e) Zones of joint exploitation. Several joint zones have been agreed between states as temporary or permanent solutions to intractable boundary disputes (see Figure 1.3.). The arrangements vary, but they generally seem to work well, deserving careful evaluation as possible solutions to boundary problems elsewhere.

(f) Maritime boundary functions. With the global system of international maritime boundaries now taking shape, analysis of their functional relationship with other types of offshore boundary at regional, national, and local levels is overdue (see Chapter 6).

The fixing of the limits of offshore state jurisdiction is only one of the political and scientific challenges associated with the oceans today.The 1982 United Nations Convention on the Law of the Sea which ushered in a new regime for the world's oceans was largely a response to the intensification of ocean use over two or more decades. It is now evident that rational planning and management of the oceans and coasts is going to be the only way to avoid increasing environmental damage and sea use conflict, especially in the 200 nautical mile exclusive economic zones and continental shelf areas. The sea is far more important to mankind than merely as a source of protein, energy, and minerals. It offers space on a crowded planet, opportunities for recreation on and below its waters, and is the means of trade and communications .between peoples. The creation of integrated regional plans for ocean management is another area in which geographers could collaborate fruitfully with colleagues in other disciplines. This theme emerged strongly during a meeting on 'New frontiers in marine geography' (convened by Adalberto Vallega and Jesse Walker) held at Barcelona in September 1986 during the IGU Regional Conference on Mediterranean Countries. This volume should encourage geographers to contribute more to the understanding and solution of maritime problems. It may also encourage marine specialists in other disciplines to welcome them as partners in an important and urgent task.

Durham

Chapter 1

WORLD MARITIME BOUNDARY DELIMITATION: THE STATE OF
PLAY

Gerald H. Blake

With the exception of Antarctica, the whole of
the earth's land surface is now effectively
partitioned between states, largely as a result of
European activity during the past century and a
half. Although land boundary disputes still commonly
occur, the trend is towards formal agreements and
progressive demarcation on the ground. By contrast,
the maritime political map of the world is in its
infancy, and it will probably take several decades
before any accurate world map of offshore boundaries
can be drawn. By the end of 1985 approximately 155
boundary agreements had been concluded between
states, concerning some 97 boundaries, fewer than
one in four of the world's potential maritime
boundaries. (1) The ratio falls even further if
boundaries associated with micro-political units are
included. For example, scrutiny of the coasts of
Brunei, Namibia, Morocco, the United Arab Emirates,
and Singapore could easily add 20 or 30 boundaries
to the total.

This chapter presents a broad summary of
progress towards maritime boundary delimitation, to
provide some context for the chapters which follow.
Analysis is confined to inter-state boundaries and
no attempt is made to discuss delimitations between
the outer limits of coastal state sovereignty and
the international seabed (or 'The Area'). This form
of maritime boundary will eventually involve
extremely complex delimitations along the outer
margins of some five million square nautical miles
of continental shelf which lie beyond 200 nautical
miles (2), where coastal states have certain
economic rights under Article 76 of the 1982 United
Nations Convention.

Discussions of offshore boundaries generally
make a point of distinguishing between territorial

1

sea boundaries, continental shelf boundaries, and boundaries between zones of extended economic jurisdiction, usually fisheries. In this chapter, at global scale, these different types of boundary, extending various distances offshore, can be largely disregarded. In recent years coastal states have increasingly sought to establish single 'maritime' boundaries for all purposes, to a maximum distance of 200 nautical miles, a trend which seems likely to continue. (Table 1.1)

Table 1.1   Types of boundary agreement 1952-1982

	Continental Shelf	'Maritime'	Total
1952-1961	2	3	5
1962-1971	20	0	20
1972-1976	22	11	33
1977-1982	9	27	36
	53	41	94

Source :   International Court of Justice, Delimitation of the Maritime Boundary in the Gulf of Maine Area, Annexes to the Reply by Canada, Volume 1 State Practice, (1983), pp 23-34.

Nevertheless the 1982 United Nations Convention maintains the distinction between types of boundary, giving different guidelines for delimitation purposes. Failing agreement to the contrary, the applicable principle for the territorial sea is the median line, unless historic title or other special circumstances dictate a different line (Article 15). Exclusive economic zone and continental shelf boundaries 'shall be effective by agreement on the basis of international law, as referred to in Article 38 of the Statute of the International Court of Justice, in order to achieve an equitable solution,' (Articles 74 (i) and 83 (i)).

In preparing Figure 1.1 single-purpose median-line boundaries were assumed wherever no boundary agreements exist. In reality many alternatives to median line boundaries are likely to emerge in future. Even with these assumptions the future pattern of world maritime political jurisdiction must invlove a measure of guesswork because of the prevalence of coastal territorial disputes and disputes over islands, sometimes involving three or more parties. In such cases a number of alternative boundary delimitations are clearly possible.

POTENTIAL BOUNDARIES BY REGION

No attempt to map the world's hypothetical
maritime boundaries at global level seems to have
been wholly satisfactory. Uncertainties over some
boundaries are compounded by cartographic problems
of scale. (3) Table 1.2 illustrates the pitfalls.
R.W. Smith does not include any world map with his
useful table, which gives a total of 376 boundaries.
J.R.V. Prescott's authoritative work (4) includes
admirably detailed regional maps from which the
totals in Table 1.2 were calculated, though not
without some difficulty. The author's figures were
derived from Figures 1.1 and 1.2 and from detailed
regional maps of the Mediterranean Sea and the
Persian-Arabian Gulf published elswhere. (5) While
the exact number of potential maritime boundaries is
not really important, the scale of the task ahead is
worth noting. Probably all the totals in Table 1.2
will prove to be rather low; out of a total of
between 376 and 400 or so potential boundaries
approximately 97 have been formally agreed, and the
International Court has given judgement on four
more. Boundary lengths were obtained for 71 agreed
boundaries and totalled 33,452 km, including the
world's longest maritime boundary (2,687 km) between
Canada and Greenland. (6) Using these as a crude
indicator, the total length of the world's maritime
boundaries is estimated to be of the order of
150,000 km.
Figure 1.1 illustrates some features of the
emerging world offshore political map. (7) Three
main types of boundary can be identified. First, the
relatively simple adjacent boundaries between states
extending to 200 nautical miles offshore, for
example around the Americas and Africa. Delimitation
here is relatively straightforward, though not
inevitably so, as the Canada - United States Gulf of
Maine case proved. Secondly, opposite boundaries,
which most commonly occur in enclosed and
semi-enclosed seas where 200 mile claims overlap.
Such delimitations, sometimes involving trilateral
negotiations, can be extremely complex. Thirdly,
boundaries between archipelagic states, notably in
the Pacific, where delimitation is complicated by
the need to draw archipelagic boundaries before
offshore claims can begin. According to J.R.V.
Prescott at least nine states in the southwest
Pacific are eligible for archipelagic baselines, but
only two have so far delimited them. (8)
In relation to its area (230,000 km^2) the

3

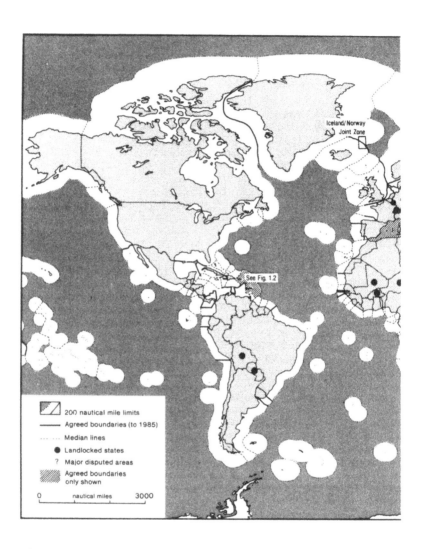

Figure 1.1. World maritime boundary delimitations in 1985 showing agreed boundaries and hypothetical median lines (Miller cylindrical projection)

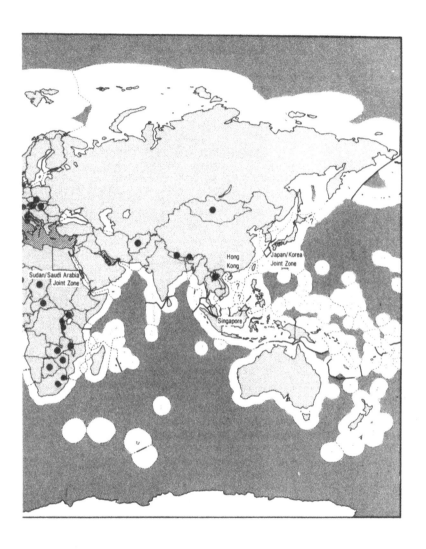

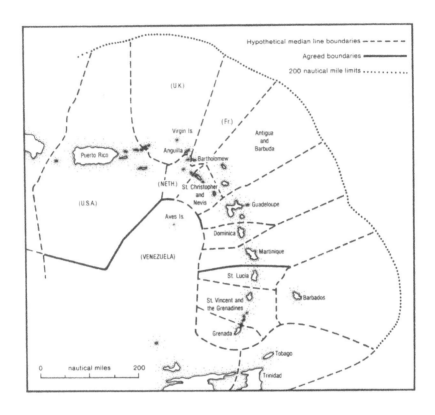

Figure 1.2. Maritime boundaries in the eastern Caribbean

Persian/Arabian Gulf probably has the highest concentration of potential maritime boundaries in the world. To date, seven out of 16 have been agreed. While offshore oil and gas have provided strong incentives to delimit boundaries for 30 years, political difficulties complicated by islands, several of them disputed, have delayed delimitation. The individual Emirates of the United Arab Emirates will eventually delimit their own offshore claims, requiring 13 more boundaries. The Caribbean and the Gulf of Mexico, (which include 33 coastal political units, 19 of which are insular) has the largest absolute number of boundaries, about

Table 1.2   Potential maritime boundaries by region

	Blake	Prescott	Smith
N.W. Atlantic	4	4	4
N.E. Atlantic/Arctic	18	24	12
S.W. Atlantic	9	9	9
S.E. Atlantic	36	35	36
Caribbean	70	81	81
North Sea	11	11	11
Baltic Sea	10	14	10
Mediterranean Sea	36	37	) 43
Black Sea	4	4	)
Indian Ocean/Red Sea	44	48	57
Persian-Arabian Gulf	16	16	16
Eastern Pacific	15	16	16
Western/ Central Pacific	48	45	51
East Asian Seas	32	30	30
World:	353	374	376

Source: R.W. Smith, 'A geographical primer to maritime boundary-making' Ocean Development and International Law Journal, 12, (1982), 1-22. J.R.V. Prescott, The Maritime Political Boundaries of the World, (Methuen, London 1985), Figures 6.2, 6.4, 7.3, 10.3, 11.2, 11.3, 12.2, 12.3, 13.2, 13.3, 13.4, 14.1.
A.D. Drysdale and G.H. Blake, The Middle East and North Africa: a Political Geography, (Oxford University Press, New York, 1985), Figures 5.2, 5.3, 5.4.

one fifth of the world total. Caribbean boundary agreements have progressed well since 1976, initiated by Cuba, Colombia, the United States and Venezuela. So far 12 boundaries out of 81 have been agreed. Future delimitations will be complicated by the island geography of the region as illustrated by Figure 1.2 showing at least 33 international boundaries in the eastern Caribbean region. Disputes between Guatemala and Belize and Colombia and Venezuela have already surfaced. Offshore resources outside the Gulf of Mexico are not great. Fisheries are small and unsophisticated, while oil and gas prospects in the central Caribbean are not bright. Certain mineral placer deposits on the seabed could be attractive in future, but there are insufficient economic inducements, so far, to create much pressure for delimitation.

A third region with a high concentration of potential boundaries is the Mediterranean Sea, where only five agreements exist, out of a possible 36 or more. Four of these are between Italy and her neighbours, and one between France and Monaco (1984). The Mediterranean exemplifies many of the world's maritime boundary problems. Complex coastal configurations and the presence of islands of contrasting sizes creates geographical and legal problems of the kind which much exercised the International Court in the Libya-Tunisia and Libya-Malta cases. In the western Mediterranean, Gibraltar and Spain's North African possessions inhibit delimitation; in the eastern Mediterranean there are questions raised by the Gaza Strip, occupied Northern Cyprus, and Britains's bases in Cyprus. Libya's claim to historic bay status for the Gulf of Sirte, and Turkey's claim to a share of continental shelf in the Aegean Sea beyond the international boundary with Greece, are added complications.

AGREEMENTS

Although at least four boundary agreements were reached worldwide before 1945, the majority have been negotiated in the last 25 years. (Table 1.3). By 1986 approximately 115 agreements had been signed, and about half a dozen more agreements await ratification and are not yet in force. Table 1.3 is based on dates when agreements entered into force. In most cases the period between signature and ratification is a year or two, but occasionally the period is much longer. The upsurge in boundary agreements in the 1970's was the result of a combination of factors. A feverish search for offshore oil followed the oil price rises of 1973-74, while the opening of the Third United Nations Conference on the Law of the Sea in 1973 gave added impetus to national concern about offshore jurisdiction. It also seems likely that a number of states waited for the important International Court Judgement in the North Sea case delivered in 1969, before attempting delimitation. The 1970s may prove to have been an exceptional decade for boundary delimitation, averaging six agreements each year. At that rate, the world offshore map would take shape in 45 years, but in reality, the process is likely to take far longer. More than half the agreements concern continental shelf boundaries, though since the mid-

Table 1.3   Boundary agreements by year

Period	Number of agreements
Before 1950	6
1950-54	2
1955-59	4
1960-64	2
1965-69	14
1970-74	29
1975-79	31
1980-85	27
Total	115

Source: R.W. Smith (ed) National Claims to Maritime Jurisdictions, Limits in the Seas 36, (U.S. Department of State, Washington, 5th Revision, 1985). S.P. Jagota Maritime Boundary, (Martinus Nijhoff, The Hague, 1985), Annex 1.

1970s the single maritime boundary has been more commonly adopted. Relatively few agreements have been exclusively concerned with territorial water boundaries and with zones of joint exploitation. (9) Of the four recent International Court Judgements, two were continental shelf cases (Libya-Tunisia 1982, and Libya-Malta 1985), and two were single-purpose maritime boundaries, (Canada-United States 1984, and Guinea-Guinea Bissau, 1983).

A highly significant feature of the world map is the emergence of zones of joint economic exploitation, generally in response to disagreements over boundary delimitation. Three contrasting examples are those agreed by Japan and the Republic of Korea for a 'joint development zone' in February 1974, by Iceland and Norway for a 'joint development zone' (south of Jan Mayen Island) in June 1982, and the Saudi Arabian/Sudanese Red Sea 'common zone' in August 1973 (Figure 1.1). The latter agreement uniquely gives the coastal states sovereign rights as far as the 1000 metre isobath, beyond which is a Common Zone where rich deposits of metalliferous muds are located. These are being exploited by a Joint Red Sea Commission. A variation on the concept of joint sovereignty and exploitation was agreed by Bahrain and Saudi Arabia as part of a boundary delimitation in February 1958. (10) A hexagonal area was put under Saudi Arabia's sovereignty, but the parties agreed to share its oil revenues. The concept of the joint economic zone could provide temporary or permanent solutions to intractable

9

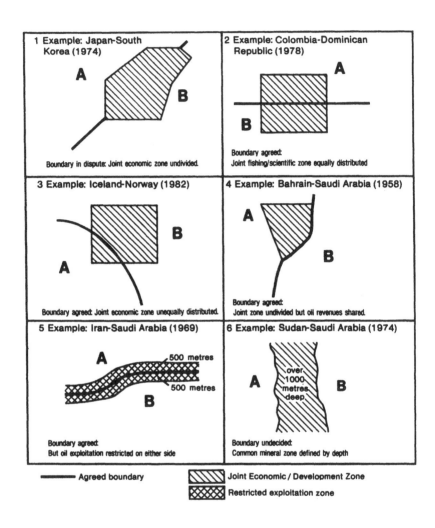

1 Example: Japan-South Korea (1974)

A

B

Boundary in dispute: Joint economic zone undivided.

2 Example: Colombia-Dominican Republic (1978)

A

B

Boundary agreed:
Joint fishing/scientific zone equally distributed

3 Example: Iceland-Norway (1982)

B

A

Boundary agreed: Joint economic zone unequally distributed.

4 Example: Bahrain-Saudi Arabia (1958)

A

B

Boundary agreed:
Joint zone undivided but oil revenues shared.

5 Example: Iran-Saudi Arabia (1969)

A

500 metres

500 metres

B

Boundary agreed:
But oil exploitation restricted on either side

6 Example: Sudan-Saudi Arabia (1974)

A

over 1000 metres deep

B

Boundary undecided:
Common mineral zone defined by depth

———— Agreed boundary

Joint Economic / Development Zone

Restricted exploitation zone

Figure 1.3. Examples of agreements between states over marine resource exploitation.

boundary disputes in other parts of the world. Several successful examples are in operation (Figure 1.3).

The regional distribution of agreements reveals few real surprises (Table 1.4). The process of boundary delimitation has clearly begun in almost every part of the world's oceans (Figure 1.1). The North Sea and Baltic possess the most complete boundary systems in the world, although some tricky problems remain in the Baltic. In certain regions (the Atlantic and Pacific fringes for example) a number of delimitations remain to be made, which should present few problems. Elsewhere (the Persian/Arabian Gulf, the East Asian seas for example) many potential boundaries are likely to be subject to serious and protracted disputes.

Table 1.4   Boundary agreements by region

Region	Potential (after Smith)	Agreed to March 1985)	Percent agreed
Atlantic/Arctic Oceans	61	18	29.5
Caribbean	81	12	14.8
North Sea/Baltic Sea	21	17	80.9
Mediterranean/Black Sea	43	6	13.9
Indian Ocean/Red Sea	57	10	17.5
Persian-Arabian Gulf	16	7	43.7
Pacific Ocean	67	15	22.4
East Asian Seas	30	7	23.2
World	376	92	24.5

Source : R.W. Smith (ed) National Claims to Maritime Jurisdictions, Limits in the Seas 36, (U.S. Department of State, Washington, 5th Revision, 1985). S.P. Jagota Maritime Boundary, (Martinus Nijhoff, The Hague, 1985), Annex 1.

DISAGREEMENTS

A very rough estimate is that two thirds of the remaining boundary delimitations might be achieved by bilateral negotiations on the basis of international law. The process may involve hard bargaining, and in certain cases possibly costly litigation lasting for months or years. The other one third are likely to be actively disputed. These disputes may be categorised as positional or territorial. Positional disputes may involve technical and legal difficulties or overlapping claims in regions of peculiar geographical complexity. Territorial disputes occur where there is no agreement as to sovereignty over islands, or on the alignment of the land boundary where it comes down to the sea. There are at least a dozen major

11

territorial disputes in coastal areas and at least two dozen active disputes over islands and groups of islands worldwide. (11) J.R.V. Prescott's thorough survey of actual and potential boundary disputes reveals that island sovereignty is the most common cause of conflict. (12) They range in size from the Kuriles (Japan-USSR) to the Tunbs and Abu Musa (Iran-United Arab Emirates). Until their ownership is agreed, maritime boundary delimitation is impossible in the vicinity of the islands.

Almost all the regions shown in Table 1.2 have anything up to five or six disputed maritime boundaries. The South China Sea with the Gulf of Thailand (2,975,000 km^2) probably possess the greatest number of potentially serious disputes. Of 15 possible boundaries, 12 are in dispute. (13) Eleven relatively powerful coastal states maintain a number of overlapping claims where hydrocarbon prospects are good, fishing is important, and where strategic considerations associated with the Strait of Malacca are very strong. The most difficult problem concerns several groups of islands including the Spratly and Paracel Islands to which there are up to four claimants, with plausible historic and geographical grounds for title. The Spratly group is claimed by China, Malaysia, Vietnam and the Philippines, all of whom have already occupied certain islands in the group. Vast areas of seabed are at stake in the central South China Sea.

It is worth remembering that coastal states have every reason to seek prompt and lasting boundary agreements. Under the 1982 United Nations Convention each coastal state is entitled to a 12 mile territorial sea and a 200 mile exclusive economic zone (EEZ) measured from the baseline from which the territorial sea is measured. By this provision, approximately 36 per cent of the ocean is due to be partitioned. Within the global EEZ is 99 per cent of the world's fish catch, (14) all the offshore oil and gas, and a valuable range of mineral resources chiefly in the form of placer deposits and aggregates of considerable commercial interest. (15) More than a quarter of the world's production is from offshore fields, and the proportion is rising. (16) Many states are also increasingly anxious to control their offshore areas for purposes of environmental management and national security. For these reasons, offshore boundary delimitation will demand a great deal of time and effort in the next decade or two.

NOTES

1. R.W. Smith, 'A geographical primer to maritime boundary-making.'" Ocean Development and International Law Journal, 12 1/2, (1982), 1-22. Earlier Smith identified 331 potential boundaries. See R.W. SMITH, 'The maritime boundaries of the United States'. Geographical Review, 81, 4, (1981), 395-410.

National Claims to Maritime Jurisdictions, Limits in the Seas 36 (US Department of State, Washington. 5th Revision 1985).

2. 1 nautical mile = 1.15 statute miles = 1853 metres R.W. Smith (ed).

3 See K. Booth. Law, Force and Diplomacy at Sea. (George Allen and Unwin, London, 1985). See also M.I. Glassner, 'The Law of the Sea,' Focus, 28, 4, (1978), 1-24.

4 J.R.V. Prescott, The Maritime Political Boundaries of the World. (Methuen : London, 1985).

5 A.D. Drysdale and G.H. Blake, The Middle East and North Africa : a Political Geography, (Oxford University Press, New York, 1985), Figures 5.2, 5.3, 5.4.

6. Boundary lengths are given in: The US Department of State Limits in the Seas series, and in International Court of Justice Libyan Counter-Memorial in the Libya-Malta Continental Shelf Case, Annex of Delimitation Agreements, Counter-Memorial, Vol II, (1983).

7. Figure 1.1 was kindly drawn by Arthur Corner of the Cartographic Unit, Department of Geography, University of Durham.

8. J.R.V. Prescott, 'Existing and potential maritime claims in the Southwest Pacific Ocean.' Ocean Yearbook 2. (Chicago University Press, Chicago, 1980) 317-345.

9. S.P. Jagota, Maritime Boundary, (Martinus Nijhoff, The Hague, 1985), Annex 1.

T.L. McDorman, K.P. Beauchamp, D.M. Johnston, Maritime Boundary Delimitation. (Lexington Books, Lexington, 1983), 157-195.

10. A. El-H. Ali, The Middle Eastern States and the Law of the Sea. (Manchester UP, Manchester, 1979), 86-91.

11. A.J. Day (ed.) Border and Territorial Disputes, (Keesing Reference Publication, Longman, Harlow, 1982).

12. J.R.V. Prescott, (1985) op.cit.

13. D.J. Dzurek, 'Boundary and resource disputes in the South China Sea'. Ocean Yearbook 5. (Chicago University Press, Chicago, 1983), 254-284.

14. J. Gulland, 'The new ocean regime: winners and losers'. Ceres, 12, 4, (1979), 19-23.

15. A.D. Couper (ed.) Times Atlas of the Oceans, (Times Books, London, 1983), 110-117.

16. E.M. Borgese and N. Ginsburg, (eds.) Ocean Yearbook 5, (Chicago University Press, Chicago, 1985), 492.

Chapter 2

DELIMITATION OF MARITIME BOUNDARIES: EMERGENT LEGAL
PRINCIPLES AND PROBLEMS

Patricia Birnie

INTRODUCTION

Recent decisions of international courts and
tribunals have provoked widely different reactions
in commentators. Some regard the outcome as
favourable-balancing certainty with flexibility,
taking account of the monotypic nature of the
problems (1) and the need for single boundaries by
concentrating on geographic factors. (2) Others are
more critical, contending that by failing to provide
precise technical guidance the courts have increased
uncertainty (3) and encouraged fragmentation rather
than uniformity as the assumed desirable aim. (4)
Some approve the outcome but find 'the ultimate
equity'(5) being left to be determined by the
subsequent conduct of the parties in relation to
it. (6)
    Some judges of the International Court of
Justice (ICJ) have accused it of embarking on the
'task of equalising the facts of nature', using
equity to achieve a compromise founded not on law,
but subjectivism, without conceptualizing equity,
indeed specifically repudiating 'over-conceptualiz-
ation'. (7)
    Even critics, however, concede that the juris-
prudence of the Court now appears fixed, (8) based
on the 1982 United Nations Law of the Sea Convention
(LOSC), (9) and recent decisions related to the
formula of achieving an equitable solution by agreement based on international law. It is questionable
whether any international tribunal could now
conclude, if a single boundary for the Exclusive
Economic Zone (EEZ) and the continental shelf is
requested, that a precise rule such as equidistance
should be applied, even though special circumstances
can be used to modify it under the 1958 Geneva

Convention on the Continental Shelf, (10) and the
Anglo-French Channel Arbitration. (11)
    To understand how far the law has moved from
the original approach it is necessary first to
outline that approach.

DEVELOPMENT AND APPLICATION OF INTERNATIONAL LAW
CONCERNING DELIMITATION UP TO 1982.

    Before 1982 the law was based on custom and
treaty: the 1945 Truman Proclamation of United
States sovereign rights to exploit the continental
shelf, leaving delimitation to be based on equitable
principles and the Continental Shelf Convention
(CSC), Article 6 of which provided that the boundary
of the continental shelves of states which are
opposite or adjacent must be determined by agreement. In the absence of agreement, unless another
boundary was justified by special circumstances, it
should be the equidistant median line for opposite
states or, for adjacent states, a line based on the
principle of equidistance from the nearest points of
the territorial sea baselines. There appeared to be
three separate methods involved - delimitation based
on agreement; on special circumstances; or on an
equidistance line, but recent cases show that
gradually this rule has been superseded.

RELEVANT CASES ON DELIMITATION: NORTH SEA CASE 1969
(12)

    The first major case to come before the ICJ,
the North Sea Cases, (12) involved two states
parties to the CSC (The Netherlands and Denmark) and
one non-party (West Germany). The Court was not
asked to delimit the area between them but to lay
down the relevant principles. It found that Article
6 of the CSC did not bind a state which was not
party to it, neither did it codify pre-existing
custom on delimitation or represent subsequent
development of customary law based upon it.
    The Court accepted neither the Dutch and Danish
arguments that equidistance was a rule of law nor
West Germany's that it was entitled to a just and
equitable share proportional to its coastline. It
found that equidistance was included in the CSC on
an experimental basis and had not crystallised into
a rule of law. The customary law was that delimitation must be concluded by negotiation in good

faith of an agreement based on equitable principles, which it distinguished from equity per se. This was not a question of applying equity simply as a matter of abstract justice but applying a rule of law which itself required the application of equitable principles, which could require use of equidistance but could also be based on other methods or combinations thereof since equidistance used alone could in some cases produce an inequitable result. The means used must be recognisably equitable, equidistance was used only when it achieved an equitable result.

The ICJ identified factors which could be taken into account in applying equitable principles but stressed that there was no legal limit to such factors, which could relate to geology or geography, such as the configuration of the coastline or the unity of deposits. The Continental Shelf doctrine emerged from recogniton of physical facts; in international law the land dominates the sea. This appeared to emphasise the concept of natural prolongation which required production of geological and geomorphological evidence.

The Court thus determined that each State should get, as far as possible, the area that was its own natural prolongation, without encroaching on another state's shelf, (13) ie the boundary was one between areas already appertaining to one or other of the states concerned. Nevertheless a reasonable degree of proportionality with the length of each state's coastline should be arrived at. Proportionality was referred to as a factor, not a legal principle, and was distinguished from according to each state a just and equal share proportional to its coastline. Equality was not required but equitable principles could be used partially to rectify inequality in a situation in which parties were otherwise in a situation of theoretical equality ie in this case, had similar lengths of coastlines fronting the same sea area. Use of equidistance or other means for this purpose would be a correct application of equitable principles; no other method of delimitation had the same practical convenience and certainty of application as equidistance, but it could produce inequity, as in this case, where a state was situated on a concave portion of an otherwise straight coastline, and often leaves to one state areas that are in fact the natural prolongation of the other. The inequity in this case was only apparent because the Netherlands and Denmark agreed bilaterally to delimit their shelves so as to box in West Germany. As the two cases were joined the Court avoided the question of whether the effect of any delimitation on a Third State should be taken into account. The Court did, however, consider whether the Netherlands and Denmark, separated as they were in the near coastal area by West Germany, could be regarded as adjacent states or whether they were more properly regarded as opposite; noting that adjacency did not require absolute proximity; it concluded they were the former.

The <u>North Sea Case</u> decision was much criticised; (14) some thought Article 6 of the CSC had become a rule of customary law; others that the definition of the continental shelf as the natural prolongation of the land territory, would open the door to geological arguments; it was more appropriate to delimitation of the outer shelf than adjacent shelves, or to non-continuous shelves; some considered that the lack of limitation of factors to be taken into account allowed for subjective inter-pretation, to cover <u>any</u> factors, such as relative population, size, or economic situation. Some judges dissented, one regretted the lack of reference to fisheries or existence of minerals and suggested that joint exploitation of unified deposits would be the best solution.

The ICJ liberated delimitation from a rigid relation to equidistance and added the choice of application of equitable principles. It did limit the latter in various ways; they must be applied as required by a rule of law, related to various factors, and be used to achieve a more equitable and reasonable result, not a just apportionment or refashioning of nature; different methods might be appropriate for different adjoining areas.

The North Sea Judgment, though intended to be limited both in legal terms and in the geographic terms of the particular case, in fact provided a number of dicta which have now been seized upon by subsequent courts to bring current doctrine and practice more into line with the formula under negotiation in the Third United Nations Law of the Sea Conference (UNCLOS) from 1973 onwards, which ultimately stressed the positive achievement of an equitable result rather than the negative achieve-ment of a reasonable mitigation of inequity.

When the next major case, the <u>Anglo French Channel Delimitation</u>, came before an international tribunal in 1977 this formula was still in process of negotiation. The Court was in no doubt that it

should take due account of the evolution of the Law
of the Sea in so far as it was relevant to the
context of that case but as there was not yet a new
Convention it was able neatly to side-step the
emerging formula and the uncertainties it
posed. (15)

THE ANGLO-FRENCH CHANNEL DELIMITATION 1977 (16)

This case differed from the North Sea one in
several ways; the Arbitral Tribunal concerned was
asked to draw the boundary; both states concerned,
France and the United Kingdom, were parties to the
1958 Geneva Convention, albeit France had made
certain reservations; the Third State affected,
Ireland, was not a party to the proceedings; there
existed in the area a small but deep trough and
several islands under United Kingdom sovereignty
situated close to the French coast.

The tribunal was asked to decide in accordance
with the rules of international law applicable in
the matter as between the parties. It was not asked
to delimit the inner coastal area. It accepted that
it should take due account of the evolution of the
Law of the Sea in so far as it was relevant to the
case, but as it was still subject to negotiation, it
did not accept that developments concerning 200 mile
exclusive economic zones (EEZ) and the continental
shelf had superseded the 1958 Convention.

It did not think that it mattered, however,
whether Article 6 of that Convention applied since
rules existed in customary international law,
identified in the North Sea case, that produced a
result similar to that laid down in the Convention.
It arrived at this conclusion by deciding that
Article 6 did not formulate the equidistance
principle and special circumstances as two separate
rules but as a single rule: a combined equitable-
special circumstances rule. It left in doubt whether
special circumstances needed to be proved in future
cases but found that the question of whether special
circumstances justified a boundary other than the
equidistance line was one of law. Thus it found
equidistance a relevant but not obligatory part of
the rule; it must determine whether in a particular
case special circumstances exist which require
deviation from it. The tribunal, therefore, felt
free to make an appreciation of the geographical and
other relevant circumstances.

It decided that the role of special circum-

stances in Article 6 was to ensure an equitable
delimitation; it did not openly start from the
premise that it must first ascertain what in its
view such a delimitation would be, though it must at
some point have made such a decision as the outcome
revealed. The tribunal stressed that the general
norm was that the boundary should be based on
equitable principles; there was no such inherent
quality in equidistance as a method as to render it
a legal norm; Article 6 used it only when geographic
and other circumstances caused it to result in an
equitable solution. Thus the rules of customary law
identified in the North Sea case were used to
interpret Article 6, though there was no evidence in
the Geneva Convention that this was its objective or
purpose; rather it had been thought that objective
clarity and precision was the aim of its parties,
not subjective evaluation of the equities, but
clearly the tribunal had now opened the door a
considerable crack to such possibilities.

The court endeavoured to keep the opening
narrow. It found that the shelf must be regarded as
a natural prolongation of the coastal state and not
encroach on another State's shelf, but that
allocating a reasonable degree of proportionality
was not a necessary criterion in all cases: states'
continental shelves must reflect the configuration
of their coasts; proportionality was a factor for
effecting a reasonable interpretation of the effects
of geography. It was <u>disproportion</u> that was the
relevant criterion rather than a general principle
of proportionality, ie proportionality was a factor
to take into account not a principle dictating a
specific result.

The court further downgraded equidistance in a
geographical context. The appropriateness of its use
was a function of relection of geography and other
circumstances; its validity in achieving an
equitable delimitation was always relevant to the
particular geographical situation; there was neither
complete freedom to use it nor no freedom to choose
other methods. A specific geographical consideration
in this context was whether the states concerned
were to be regarded as opposite or adjacent to each
other since equidistance often produces a more
distorting effect in the latter than the former
situation. The court in this case treated the two
states as being in different situations in the
Channel (opposite) and in the Western Approaches
area (adjacent) after an examination of the
geographical facts and their geographical relation

to each other. The Hurd Deep was discounted as too
insignificant a geographical feature to exercise any
material influence on the boundary. The shelf in the
area was found to be essentially continuous, as in
all other cases under review in this chapter, though
the court did not find that such continuity required
uniform treated of delimitation along all points of
the lines to be drawn.

The court evolved an unusual approach, which
has subsequently become usual in such cases, to the
problem created by the fact that British islands
were situated so close to the French coasts on what
would otherwise be the natural prolongation of
French territory. It designated this situation as
placing the islands on the 'wrong side' of the
Channel median line and proceeded to evaluate the
region as a whole. It found that the islands
disturbed the balance of the goegraphical circumstances in a region where states had a broad
equality of coastlines; some account should be taken
of this in balancing the equities, though equity did
not require equality. The weight it gave the islands
was attributed to their relative size and importance
and their existing territorial sea and 12 mile
fishing zone. In according them a 12 mile
continental shelf within the median line, the court
disengaged itself also from a rigid interpretation
of natural prolongation: it was neither to be set
aside nor regarded as absolute. In such a situation
it was a judicial concept susceptible of legal
qualification: it certainly did not require that all
the Channel Islands Shelf be attributed to the
United Kingdom. It took over only circumstances and
was qualified by the customary law which required
application of equitable principles; it could
therefore be modified by special circumstances. The
islands' situation was a circumstance creative of
inequity for France since they were wholly detached
from the United Kingdom and not a coastal fringe or
enclave of the latter.

The court dealt with the drawing of the Western
boundary by distinguishing an adjacent rather than
opposite relation between the two states. It found
that the Scilly Isles, in the region as a whole,
disproportionately pushed an equidistance line in
the direction of the French coast and decided that
equity required an abatement of the effect; though
it did not require, and could not achieve, a
refashioning of nature. The answer was not entirely
to reject equidistance but to modify it by giving
less than full effect - in this case half effect -

to the Scillies. It found that it was not required to take any account of the effect of this on future delimitations with Ireland.

Once again a court had opted for a reaonable compromise which, despite careful qualifications in terms of legal principles and geographical circumstances, opened the door to some element of subjective evaluation of the equities and of choice of means for observing them. Nonetheless, the decision was welcomed as a common sense approach, toning down the geological arguments deriving from the North Sea case decision: natural prolongation was not seen to be of major relevance in delimitation between opposite and adjacent states; proportionality was not regarded as a principle but a mitigating factor. (17) The legal limits to giving effect to equitable principles were, however, left vague and, in the view of some, special circumstances were too closely related to them. The wide view the court took of the area subject to delimitation also proved to be significant.

Each case presents a different geographical situation, and each court is differently constituted. When the next relevant case came before the ICJ in 1982, (18) its membership was considerably changed. This no doubt affects the outcome as the legal background and perspective of the judges is re-oriented. The question put to each court is also different and narrows or enlarges the scope of the decision. Professor Briggs had foreseen in the Anglo-French case that 'the rules of positive law in Article 6 will be eroded by its identification with equitable principles, permitting attempts by the court to redress the inequities of geography' (19) since the burden of proof of special circumstances had been shifted from the state invoking them to the Court itself. As others noticed 'the concept of equitable principles is a dangerous one if applied loosely, given its high degree of abstraction'. (20)

THE TUNISIAN-LIBYAN ARAB JAMAHARIYA CASE 1982 (21)

Libya and Tunisia asked the IJC to decide what principles and rules of international law may be applied to delimitation of the shelf between them and to take its decision according to equitable principles, the relevant circumstances which characterise the area, as well as the new accepted trends in the Third Conference on the Law of the

Sea.

There was now more state practice on delimitation and the UNCLOS III was considering a draft Convention (22) that defined the continental shelf both in terms of natural prolongation and 200 mile distance from the baselines. The ICJ regarded this as a two-part definition subject to different criteria, and concluded that the legal concept of the shelf solely as a species of platform had been modified; the first part was regarded as irrelevant to the case for reasons which will be referred to later. By 1981 the LOSC delimitation formula had two versions: one referring to agreement in accordance with equitable principles, employing the median or equidistance line, and taking into account all circumstances prevailing in the area concerned; the other, newer text, to agreement on the basis of international law, as referred to in Article 38 of the Statute of the International Court of Justice, in order to achieve an equitable solution. The ICJ noted that the newer text indicated no specific criteria to guide parties to an equitable solution, but emphasised the kind of solution to be achieved.

Libya based its case on elaborate geological arguments, based on natural prolongation. The court dismissed them, finding that it must examine equitable principles divorced from 'the concept' of natural prolongation. It concluded that the result of equitable principles must be equitable, though characterisation of equitable in terms of both means and ends was unhelpful. It added:

> It is, however the result which is pre-dominant, the principles are subordinate to the goal. The equitableness of a principle must be assessed in the light of its usefulness for the purpose of arriving at an equitable result. It is not every principle which is in itself equitable; it may acquire this quality by reference to the equitableness of the solution. The principles to be indicated by the Court have to be selected according to their appropriateness for reaching an equitable result. The term 'equitable principles' cannot be interpreted in the abstract. It refers back to the principles and rules which may be appropriate in order to achieve that end. (23)

It based this conclusion on a dicta in the

North Sea case that the problem was one of defining the means whereby the delimitation can be carried out in such a way as to be recognised as equitable, despite the fact that the Court had stressed that it was equity as a rule of law that itself required the application of these principles. The Court confused the legal principles with the methods and factors to be used to fulfil them in an equitable manner, but it said it was:

> bound to apply equitable principles as part of international law and to balance up the various consideration it regards a relevant in order to produce an equitable result. While it is clear that no rigid rules exist as to the exact weight to be attached to each element in the case, this is very far from being an exercise of discretion ... (24)

Very fine lines were being drawn; so fine as to be invisible to some.

The court dismissed both Tunisia's and Libya's geological arguments and evidence since the shelf in this case was continuous and the natural pro-longation of both states; geologists might still retain a role if any future dispute concerns an interrupted shelf. The court then had to decide what were the circumstances relevant to producing an equitable result; they might include geography (coastal formation, excluding overlapping shelves); presence of islands; historic fishing rights, which Tunisia argued could not be left 'on the wrong side of the line'; or other economic factors - Tunisia argued that compared to Libya it was a poor state deprived of resources.

The court took account of geography, gave some effect to the change in direction of the Tunisian coast, gave half effect to some islands (the Kerkennahs) but unaccountably ignored others; found that as the fisheries were not affected by the boundary it laid down, the question of their relevance did not arise (leaving it open for their relevance to be argued in other cases) and rejected Tunisia's future economic development. It thus introduced the idea that factors conducive of certainty are to be preferred, an argument seized on later by the ICJ in the Gulf of Maine Case.

Equidistance was thus further downgraded, on its way to being discarded as a principle and reduced to a method amongst others. The court found it was not required even to examine it as a first step. The balancing up of relevant circumstances must precede resort to it: it was neither a mandatory principle nor a method having privileged status against others. There was no single obligatory method in international law; several methods could be applied to the same delimitation, starting with the particular geographical situation.

For this purpose the court defined the area it considered relevant to the case; the specific areas could be treated differently in relation to achieving an overall equitable result. It questionably accorded special significance to a pre-existing informal de facto line dividing oil concessions as indicating a line the parties regarded as equitable. It determined that the most evident geograpical feature to be taken into account was a radical change in the general direction of the coast. It did not take into account the repercussions on Italy and Malta, despite its rejection of Malta's attempt to intervene. (25) Finally, to test whether the decision met the requirement of proportionality as an aspect of equity, the Court found that the relevant coastline of Libya stood in a relation of 31:69 to that of Tunisia and that, taking account of all the relevant circumstances, the result met the test.

Some have criticised the decision as equating equity with equality; others consider it a significant step towards formulation of integrated principles applicable to delimitation of customary maritime boundaries for both the EEZ and continental shelf by reducing the significance of the geological factor and applying a very general formula which allowed for consideration of all relevant circumstances, within the general geographic relationship of the parties, respecting natural prolongation of coastal fronts only, to avoid cutting off one state from the areas off its coast, the principle of non-encroachment being respected. (26) There were dissenting opinions and it remained unclear whether the stress on the result as the test of the equitableness of a principle was a correct view of what is required by equity. (27)

THE GULF OF MAINE CASE 1984 (28)

This case is the subject of Chapter 12 and has been extensively analysed elsewhere; (29) it will suffice here to pick out emergent trends and problems.

There were several novel features. It was the first case involving two parties to the 1958 Convention to come before a Panel of the ICJ and specifically to request it to draw the boundary; the first to request a single _ boundary for the continental shelf and an exclusive fisheries or economic zone, although Australia and Papua New Guinea had, by agreement, established separate seabed and fisheries jurisdiction lines, with a protected area and arrangements for sharing the fish catches. (30) The United States and Canada disregarded this solution, having been unable to negotiate an acceptable fisheries agreement for the Gulf -, perhaps because they had not negotiated in the wider context and had divorced fisheries from other aspects of delimitation, perhaps because they insisted on a single method of delimitation, whereas in fact no single <u>method</u> is suitable for universal application, only the single <u>goal</u> of equitable delimitation. (31) By asking now for a single boundary the parties limited the court's scope to fashion wider solutions of this kind. (32) The Panel found international law permitted it to draw a single boundary and concluded that:

> ... when agreement cannot be achieved, delimitation should be effected by recourse to a Third Party possessing the necessary competence...delimitation is to be effected by the application of equitable criteria and by the use of practical methods capable of ensuring with regard to the geographic configuration of the area and other relevant circumstances, an equitable result. (33)

This further disengaged it from legalistic ideas concerning application of equitable principles and the rule of equidistance mitigated by special circumstances developed in <u>North Sea</u> and <u>Anglo-French Channel Cases</u>, in their place substituting further indeterminate equitable criteria, use of unidentified 'practical methods', and a stress on the geographic configuration of the coastline. The choice of method, said the court, was to be determined primarily by geography, since that could be said to have a 'neutral character'. Geographers may prefer to say that geographic features are merely unbiased, since they may in fact be disadvantageous to one state. The Panel rejected

United States arguments that coastal fronts could be categorised as of primary and secondary importance for delimitation purposes.

To fulfil its task the court found it necessary to define the Gulf of Maine, alleging that no adequate definition existed. It did so in an arbitrary geographical fashion, treating it as a rectangle. It stressed the monotypic nature of each case and the need to avoid attempts to conceptualize the application of the principles and rules concerning the continental shelf; it downplayed these, emphasising geographic characterisation.

The court used different methods for different sections based on different geographic criteria (adjacent; opposite; outside the Gulf area); political and economic considerations were rejected. Effect on fishing was taken into consideration only as a test of the equitableness of the result. As no catastrophic repercussions were envisaged, each State allegedly having retained the fishing most important to it, no further adjustment was made. Unlike the Libya-Tunisia de facto concession lines, here the Court did not regard the interim fisheries agreements as binding on the parties in relation to the final boundary. Economic factors, such as effect on the coastal community and the fishing management implications, were given scant attention; evidence concerning ecological and environmental factors - such as currents, water masses and fish patterns, were dismissed as being too uncertain and variable, despite the fact that a fishery **and** other purpose boundary was involved; these were not treated as neutral criteria like geography.

The court distinguished between the principles and rules and the equitable criteria or practical methods used to ensure that a particular situation was dealt with in accordance with the rules; customary law could only provide a few basic legal principles to serve as guidelines and could not be expected to specify the criteria or methods. It found that the relevant provisions in the 1982 LOSC (Articles 73 and 84) had been adopted without objection and were consonant with the general international law on the question that required conformity to equitable principles, having regard to all relevant circumstances, to achieve an equitable result. It rejected all the methods proposed by the parties and put forward its own independent solution derived from geography; it found that only geometric methods would serve its purpose of avoiding unreasonable effects and producing an equitable solution. In the three segments it contrived, it took account, <u>inter alia</u>, of coastal configuration, gave half effect to one island and drew a perpendicular out to sea on an artificially constructed so-called 'closing line' of the Gulf. Finally, it looked at the effect on the subsistence and economic development of the population concerned to verify whether the result could be considered intrinsically equitable in the light of all the circumstances, though it did not find such verification to be absolutely necessary and found the effects to be inconsequential. It relied on the long tradition of friendly and fruitful co-operation in maritime matters between Canada and the United States to surmount any difficulties.

There were dissenting judgements: Judge Gros disputed the legality of a single boundary, thought the judgement relied too heavily on the UNCLOS III's work, and concluded tht the Chamber's reasoning implied that there was now no legal rule governing maritime delimitation since the principles relied on, methods used to effect them, and connections made, transformed the whole operation into an exercise wherein it would henceforth be open to each judge to decide at his discretion what is equitable. (34)

Neither side realised the full implications of asking for a single boundary which was not <u>required</u> by any rule of law, or properly addressed the question of whether in such cases the economic zone or continental shelf criteria should predominate or whether a new approach was required. The Torres Strait solution certainly provided for more flexibility. In opting for the certainty of geographical criteria, the court may have produced a less equitable solution than it might ' have. The results may not be catastrophic for United States fishermen, but have not been welcomed by them. The United States made a case for the relevance of environmental factors and still lost; though the court did not close the door to their admission in some future case, it is difficult to imagine that a stronger case could ever have been made.

The case inaugurates a trend to establishment of single maritime boundaries for both the EEZ and contintental shelf; international law does not forbid this. The trend to discount geological and environmental arguments in favour of geographical factors continues. Other factors have not been judged irrelevant but it is clear that anything other than the most dire economic effects on local populations are irrelevant. The court will, if the parties do not define the area, _itself_ do so in the manner _it_ judges relevant, and will resort to whatever practical methods of delimitation _it_ judges necessary, to achieve whatever equitable solution _it_ has in mind for that geographic configuration, after balancing the factors of _its_ choice, though it still regards its choice as restricted by equitable principles. The case was heard by only a five member panel of the ICJ, the choice of which was influenced by the parties; possibly a full Court might decide differently. It is submitted, however, that the judgement continues the trend to disengagement from strict application of equitable principles to permit a free appreciation of factors to achieve an equitable solution determined by such factors. The relevant factors are, however, gradually being clarified, especially in the following cases.

THE MALTA-LIBYA CASE 1985 (35)

Though the attempt of Italy, the state most affected by the outcome of this case, to intervene was rejected by the ICJ on jurisdictional grounds, the court did not ignore the effect on Italy and the case is particularly interesting in that respect. The ICJ found that it could not completely ignore Italy's legal interest and though inhibited by the question put to it, which excluded consideration of the effect on Third States, did refer to them since in effect the whole area was in dispute, not just the area in issue. The Tribunal in the Guinea-Guinea Bissau Case had taken a similar view as we shall see.

This case is also interesting in that the court for the first time accepted that the concept of a 200 mile continental shelf approved in the LOSC 1982 was now as much part of customary international law as that based on natural prolongation. Thus within 200 miles, whether or not the shelf was continuous, each state is entitled to a continental shelf; geological evidence has become irrelevant except in relation to establishing rights beyond 200 miles and outer limits. Libya had argued that natural pro- longation was the fundamental basis of title; Malta had favoured the distance criterion. The EEZ or Exclusive Fishing Zone (EFZ) concept had been accepted in the Maine Case; the Court now found that the seabed rights involved were inevitably linked to the continental shelf concept, outside as well as

29

inside the LOSC, despite the fact that the LOSC does not automatically confer EEZ rights on coastal states. It is generally thought that states must claim their EEZ, although shelf rights exist _ipso facto_. Canada and the United States had claimed only EFZs when they brought the case, though in 1983 the United States declared an EEZ, (in terms different from the LOSC). The ICJ in the Libya-Malta case referred only to rights which coastal states may proclaim, but it seems they will now have continental shelves in terms of the LOSC Article 76 definition thrust upon them whether or not they are parties to that convention or have claimed an EEZ: the court found that the permissible EEZ is now a relevant circumstance.

It still remains to delimit such areas. Malta argued that though equidistance was not an obligatory method of delimitation for shelves based on distance it was the appropriate starting point, though it required modification in this case. Neither Article 74 nor 83 of the LOSC refer to either this principle or equitable principles. The court found that the numerous delimitation agreements now concluded did not support Malta, though equidistance was still used when it produced an equitable result.

The court was asked not to draw the line, but to lay down the applicable principles and indicate how to apply them in a practical fashion. It defined for itself the area in issue, taking a macro-geographic perspective of the whole Mediterranean area, particularly noting Malta's isolation in a vast area. It identified the median line between Malta and Libya, and also, curiously, the line that would result if Malta was an Italian possession but Italy received no share based on its sovereignty over Malta. It then took the difference between the two lines based on these hypotheses, drew a line equidistant from the Malta-Libya coasts and modified it northwards. This it considered was an application of equitable principles leading to an equitable result which took account of the relevant circumstances - the disparity between the length of Malta's coast (24 miles long) and Libya's (192 miles long) - and the legally permissible extent of the EEZ.

It rejected both Libya's argument that a state with a greater land mass should have a larger share of the shelf, and Malta's that the relative economic position of the two states should be taken into account and that their maritime areas be of equal juridical value.

The court specifically identified some equitable principles which it said are applicable also in delimitation by agreement, viz:

(i) There could be no question of refashioning geography, or compensating for the inequalities of nature;

(ii) Non-encroachment by one party on the areas appertaining the other;

(iii) The respect due to all relevant circumstances;

(iv) Equity did not necessarily imply equality or make equal what is unequal, though all states are equal before the law;

(v) There could be no question of distributive justice. (36)

These are all principles applied in previous cases. The ICJ found that application of such principles constrained the number and choice of relevant circumstances, though there was no legal limit to them, since only principles pertinent to the institution of the continental shelf as it has developed within the law and to application of these principles to its delimitation, qualified for inclusion, not 'considerations strange to its nature'. (37) The choice of principles and of criteria will vary depending on the particular geographic situations in a given case.

The case brought together a number of specific equitable principles and identified a new relevant circumstance: the 200 mile zone of continental shelf; since the LOSC defines seabed rights in the zone by reference to the latter regime, there can be a continental shelf without an EEZ but the vice versa situation is impossible. This makes geological characteristics of natural prolongation superfluous; no role need now be ascribed to geological or geophysical factors in the context of the first 200 miles of 'legal' shelf.

THE GUINEA-GUINEA BISSAU CASE 1983 (38)

This case came before a three-member arbitral tribunal which was asked to decide whether a convention of 1886 between Portugal and France had determined the maritime boundary between these states and, if not, to delimit the boundary of the territorial sea, EEZ and continental shelf by a single line. This they proceeded to do, having answered the question in the negative.

Following the Gulf of Maine case, the Tribunal

found that international law provided only certain legal principles, indicating the factors to be used to achieve an equitable solution. The choice of factors and methods lay with the tribunal, and though emerging from the legal rules were also factors laid down by physics, mathematics, history, politics and economics. These factors were not related; none was <u>required</u> to be used: each delimitation was 'unicum'. (39) The factors, therefore, had to be derived from the case, especially the characteristics peculiar to the region, as judged by the Tribunal to be relevant.

The court defined the region; it took a broad perspective of that region, assessing the configuration of the coast, not just from the frontiers of the two Guineas but from what it called a 'global' viewpoint - from the outermost points on the coasts of the neighbouring states of Senegal and Sierra Leone - to determine more accurately the general direction of the coast and to enable the delimitation to fit into actual and future delimitations in that region of West Africa. It stressed both the need for simplicity in tracing the coast and also in the drawing of the line; whilst its decision had to fit its sense of justice and be based on reason grounded in law, it must operate on simple lines, adapted not only to being precisely drawn on the maps actually to be used but also to the techniques available to those who would use them.

Looking at the circumstances the Tribunal found the shelf was continuous: geological evidence adduced to rebut this assumption was dismissed as too vague and uncertain. It assessed the relative weight to be accorded to islands off the coast and analysed them under three different categories. Some it used roughly to compensate Guinea's shorter coastline so that both states could be regarded as having the same 154 mile coast; one small Guinean island on the high seas, on the 'wrong side' of the line drawn, was allowed to retain its territorial sea. The Tribunal insisted that the delimitation must leave to each state the islands under its sovereignty.

The parties agreed that nature could not be refashioned and that parts of the shelf which properly appertained to one state should not be cut off from it or enclaved, but they had disagreed on whether equidistance was the appropriate starting point, to be modified by circumstances, or whether any one practical method should be given priority.

The court, following the trend, concluded that equidistance was only a method comparable to others and that even if it did have some intrinsic quality because of its scientific character and ease of application, it should not be given priority, but used only if it achieved the objective of the equitable result. It re-iterated that the method to use was determined by the result and the consequent judicial reasoning. Equidistance was inappropriate to the concave nature of the West African coast as it increased the risk of cutting states off from their rightful offshore areas. Casting an eye over the whole region and taking an expansive view of its coastal formation and direction, the court drew a line that was a compromise between the most extreme position of both states.

It verified the equitableness of this, _ex post facto_, by reference to the continuous nature of the shelf in the region, which was a natural prolongation of both states (ignoring the physical factors of geology, which were little known) and the test of proportionality between the two areas, related to the length of coastline, not to land mass, as evidencing judicial not mathematical equality since it was not a mechanical rule. Economic circumstances were not taken into account as they were impermanent and a Tribunal could not compensate economic inequality by adjusting a delimitation line dictated by objective considerations. The parties should co-operate over resource management in order the better to realise their legitimate economic and development needs. The trends were being closely followed.

CONCLUSION

The major change of emphasis in the international law concerning delimitation is apparent in the three most recent cases on the subject. The formula expressed solely in terms of achieving the objective of an equitable result allows judges considerable freedom to choose both the relevant principles and relevant factors, circumstances or criteria, though there seems to be some confusion between these terms. The courts have, however, identified a number of the relevant equitable principles and narrowed down the relevant circumstances, both by taking account of some, such as geographic configuration of the coast and the presence of islands, and rejected others, such as the local economic, security or political aspects, though these sometimes enter into the verification process; proportionality and non-encroachment are regarded as safeguarding these interests as well as other equities.

It seems that the Courts, including the ICJ itself, have brought themselves into line with the 1982 LOSC so that their judgments will not be rendered obsolete or inequitable in comparison with any given on the basis of that Convention once it enters into force. It has been suggested that looking even further ahead, the ICJ is ensuring that it stays competitive with the future Law of the Sea Tribunal in the market for further delimitation cases. In closing one door to geologists and environmental scientists it has, however, opened up another to the welcome visitation of geographers to help it in the delicate tasks of fashioning regions, coastal lengths and configurations, without actually refashioning nature. As recent commentators succinctly stated: 'Geography is not so much the most neutral factor as the most positive factor in relation to both the seabed and the water column'. (40)

NOTES

1. K. Troy 'The making of offshore boundaries: beyond the Gulf of Maine', Part I, Oil and Gas Law and Taxation Review, 11, (1984), 289-298, Part 2, 12, (1985), 314-328.

2. T.L. McDorman, P.M. Saunders, and D.L. Vanderzwaag, 'The Gulf of Maine boundary: dropping anchor or setting a course?', Marine Policy, 9, (1985), 90-107, 107.

3. A.E. Chircop, and I.T. Gault, 'The Making of an offshore boundary: The Gulf of Maine case, 1984', Oil and Gas Law and Taxation Review, 7, (1984/85), 173-181, 173.

4. Ibid. 180-181.

5. M.B. Feldman, 'The Tunisia-Libya Continental Shelf case: geographic justice or judicial Compromise?' American Journal of International Law, 77, (1983), 219-238, 238.

6. J. Schneider, 'The First ICJ Chamber experiment: the Gulf of Maine Case: the nature of an equitable result', American Journal of International Law, 79, (1985), 539-577, 575-577.

7. Dissenting Opinion of Judge Gros in Case Concerning the Continental Shelf (Tunisia/Libyan Arab Jamahiriya), ICJ Report, 1982, XXI International Legal Materials, 1982, 143-156, at 149-155; ibid., in Case Concerning Delimitation of the Maritime Boundary in the Gulf of Maine Area, ICJ Report 1984, 360-389, 377.

8. Gulf of Maine Case, Ibid., 379.

9. United Nations Convention on the Law of the Sea 1982, (United Nations, New York, 1983).

10. United Nations Treaty Series, 499, 311.

11. Arbitration between the United Kingdom of Great Britain and Northern Ireland and the French Republic on the Delimitation of the Continental Shelf 1977, Cmnd. 743, (HMSO, London, 1977).

12. North Sea Cases, ICJ Report, 1969, 1.

13. For a discussion of the subsequent development of this concept, generated by this case, see D.N. Hutchison, 'The Concept of National Prolongation in the Jurisprudence Concerning Delimitation of Continental Shelf Areas', British Yearbook of International Law, 57 (1984), 133-187. He concludes that there are now seven different senses in which the term may be found in the current jurisprudence, each performing a different normative role; at 184-5. As illustrated in this chapter of this book, natural prolongation is, however, no longer emphasised as a general or dominant principle, nor in its geological aspect. It is now relegated more to geographical aspects and used to test the equitableness of the final delimitation.

14. See for example W. Friedman, 'The North Sea Continental Shelf cases - a critique', American Journal of International Law, 61, (1970), 229-240; E.D. Brown, The Legal Regime of Hydrospace (Stevens, London, 1971).

15. E.D. Brown, 'The Anglo-French Continental Shelf Case', San Diego Law Review 6, (1979), 461-530.

16. Arbitration between the United Kingdom of Great Britain and Northern Ireland and the French Republic on the Delimitation of the Continental Shelf 1977, Cmnd. 743, (HMSO, London, 1977).

17. E.D. Brown, 'The Anglo-French Continental Shelf Case', San Diego Law Review, 6 (1979), 461-530.

18. Tunisia/Libya Arab Jamahiriya Case; op.cit., (see note 7).

19. United Nations Treaty Series, 499, 126.

20. L. Herman, 'The Court Giveth and the court taketh away: an analysis of the Tunisian-Libyan Continental Shelf case', International and Comparative Law Quarterly, 33 (1984) 825-858, 842.

21. For contrasting views of this case see E.D. Brown, 'The Tunisian-Libyan Continental Shelf case: a missed opportunity', Marine Policy 7 (1983), 142; M.B. Feldman, 'The Tunisian-Libyan Continental Shelf Case: Geographic justice or judicial compromise?' American Journal of International Law 77 (1983), 219.

22. Draft Convention on the Law of the Sea, UN Doc. A/Conf.62 L78, (1981).

23. Libya/Tunisia Case Judgment, para.70.

24. Ibid., para.71, emphasis added.

25. Ibid., para.10.

26. Feldman, op.cit., 220, (see note 19).

27. Herman, op.cit., 857, (see note 19), suggests that the whole process of evaluation of complex factors in order to render justice in the particular case was not the best approach to application of equitable principles.

28. Case Concerning Delimitation of the Maritime Boundary in the Gulf of Maine Area, (1982), ICJ Report 246.

29. See, for example, the articles cited in notes 1, 2, 3 and 5.

30. For details of their novel solution and analysis thereof, see H. Burmeister, 'The Torres Strait Treaty: ocean boundary delimitation by Agreement', American Journal of International Law, 76, (1982), 321-349; for hints of a similar solution for the Jan Mayen Area; see R. Churchill, 'Maritime Delimitation in the Jan Mayen Area', Marine Policy, 9, (1985) 1, and W. Oestreng, 'Regional delimitation agreements in the Arctic Seas: cases of procedure?', Paper presented at the 18th Annual Conference of the Law of the Sea Institute, San Francisco, USA, 1984. For the 1984 Iceland-Norway Agreement on the Continental Shelf between Iceland and Jan Mayen, see, International Legal Materials, XXI (1982), 1222.

31. Burmeister, op.cit., 349, (see note 30).

32. Sung-Myon Rhee, 'Equitable solutions to the maritime boundary disputes between the United States and Canada on the Gulf of Maine', American Journal of International Law, 75, (1981).

33. Gulf of Maine Case, op.cit., para.240, see note 7.

34. Ibid.,360-389.

35. Case Concerning the Continental Shelf: Libyan-Arab Jamahariya, Malta 1985, ICJ Report 13, (1985).

36. Ibid., para.46.

37. Ibid., para.49.

38. 'Tribunal Arbitral Pour la Délimitation de la Frontière Maritime Guinee/Guinee-Bissau, 14 Février 1985', Revue Générale de Droit International Public, 89, (1985), 484-537; for a description and analysis of the case see K. Troy, 'The making of offshore boundaries: beyond the Gulf of Maine', Part II; Oil and Gas Law Taxation Review, 3 (1985) 314-328.

39. Guinea-Guinea Bissau Judgment, (1985) para.89.

40. L.H. Legault and Blair Hankey, 'From sea to seabed: the single Maritime Boundary in the Gulf of Maine case', American Journal of International Law, 79, (1985), 961-991, 990.

Chapter 3

STRAIGHT AND ARCHIPELAGIC BASELINES

J.R.V. Prescott

> The Court does not express any opinion on
> whether the inclusion of Filfla in the
> Maltese baseline was legally justified:
> ... (1)

That must have been a matter of relief to the
Maltese authorities because any objective investigation must lead to the conclusion that the
island of Filfla cannot be tied into a system of
straight baselines according to either the 1958
Convention of the Territorial Sea and Contiguous
Zone or the 1982 Convention on the Law of the Sea.
If Malta had been pilloried for breaking the rules
dealing with baseline delimitation it could have
pointed to many other countries which have breached
the rules in a much more blatant fashion. Because
the rules covering baselines in the 1958 and 1982
Conventions are ambiguous and because there is no
international authority charged with their
supervision there has been widespread abuse of the
system. Indeed it would now be possible to draw a
straight baseline along any section of coast in the
world and cite an existing straight baseline as a
precedent.
    Both the 1958 and 1982 Conventions describe the
low-water line as the 'normal' baseline. Although
closing lines across the mouths of rivers and bays,
straight baselines along indented coasts or coasts
fringed with islands, and archipelagic baselines
surrounding archipelagic states were not described
as abnormal baselines, there was presumably an
expectation that in a global sense such straight
lines would be exceptional. In fact many countries
now measure their maritime claims from straight
lines even though the coast concerned has none of
the features which justify departure from the low-

water line.

The original intention of substituting straight lines for the low-water mark was to avoid situations where the territorial waters are penetrated by deep corridors of non-territorial waters or surround enclaves of such waters. The desire was always to simplify the alignment of the outer limit of the territorial seas. Such simplification was of benefit to innocent alien navigators and surveillance authorities of the coastal state. It seems probable that the unjustified use of straight lines is primarily designed to increase the width of the combined zone of internal and territorial waters for security purposes. States may also use such lines to gain an advantage in negotiating common boundaries with neighbouring states. Only when individual legs of the baseline system are long can the coastal state hope to extend its exclusive economic zone.

There is a fundamental distinction between closing lines on the one hand and straight and archipelagic baselines on the other. Closing lines deal with single features and they are generally short. The exceptions provided by the improper closure of the gulf at the mouth of the Rio de la Plata by Argentina and Uruguay and of the Khalij Surt (or Gulf of Sirte) by Libya can be discounted. (2) In contrast straight and archipelagic baselines deal with multiple features and may extend over long distances. Long sections of the coasts of Norway, Chile and Canada provide models for situations justifying the construction of straight baselines and the 1982 Convention specifies a maximum length of 125 nautical miles for any individual leg of archipelagic baselines. This analysis deals with the regional scope of straight and archipelagic baselines rather than the local effect of closing lines.

There are important differences between straight and archipelagic baselines. The concept of straight baselines has a long history and it is thoroughly accepted as a legitimate basis for action by states. The rules for drawing straight baselines are notoriously ambiguous and they have been broken in many ways by several states. Finally straight baselines are common around the shores of all oceans except the Arctic Ocean. The concept of archipelagic baselines is as recent as the 1982 Convention which is not yet in force. Only comparatively few countries have drawn such baselines and none has breached the apparently precise rules in an objectionable manner.

STRAIGHT BASELINES

Article 7 of the 1982 Convention deals with straight baselines. Three of its six paragraphs were copied exactly from the 1958 Convention and two others from the same source were only slightly altered. One completely new paragraph was added dealing with unstable coasts.

This chapter describes three types of coast where straight baselines may be employed. First there are those coasts which are 'deeply indented and cut into' and second there are situations where there is 'a fringe of islands along the coast in its immediate vicinity'. (3) These twin concepts and the language which expresses them can be traced from the judgement of the International Court of Justice in the case between Norway and the United Kingdom in 1951 to the report of the International Law Commission in 1956 to the 1958 Convention. (4) Had these concepts remained true to their birth on the deeply indented fjord coast of northern Norway which is fringed by the countless thousands of isles and rocks in the skjaegaard the issue of straight baselines would have been unexceptionable. Instead the concepts have been distorted beyond recognition by increasingly liberal interpretations of the words which framed them.

The third type of coast was included only in the negotiations leading to the 1958 Convention. States are now empowered to draw straight baselines along the furthest seaward extent of the low-water line of deltaic coasts which are 'highly un-stable'. (5) Then if the coastline retreats the straight baseline can remain anchored in the sea and maritime claims will continue to be measured from it.

Article 7 lays down five general principles which should govern the construction of straight baselines along appropriate coasts. First the line should not depart to any appreciable extent from the general direction of the coast. The International Court of Justice which originated this principle noted that it 'is devoid of any mathematical precision'. (6) Hodgson and Alexander discovered that apart from a solitary exception the Norwegian straight baseline did not deviate from the general direction of the coast by more than 15°. (7) Unfortunately different geographers might justify using different lengths of coast to determine its general direction. The Court could only offer the advice that except in cases of manifest abuse it was unsatisfactory to examine one sector alone or to rely on impressions gained from large scale charts.

The second principle is that waters lying landwards of the straight baseline should be sufficiently closely linked to the land domain to be subject to the regime of internal waters. The International Court of Justice noted that this concept underlay the closure of bays, but argued that it should be more liberally applied along coasts with the unusual configuration of northern Norway. (8) This call for a more liberal application has been followed with remarkable enthusiasm by countries such as Burma and Vietnam.

The third principle prohibits the use of low-tide elevations as basepoints for a straight baseline unless they are surmounted by a lighthouse or similar installation or if their use for this purpose has received general international recognition. This last qualification was included to meet the case of Norway where two of its basepoints on the 1935 line approved by the Court were on low-tide elevations. Unfortunately its inclusion provides a loophole which can be exploited. A country could now announce its baseline and use some low-tide elevations and then announce after a few years that the absence or low level of criticism constitutes evidence of general international recognition.

When straight baselines are deemed to be appropriate the coastal state may take into account important economic interests of long-standing. This is a vague prescription but there is no evidence that any country has determined the alignment of any section of its straight baseline according to this principle.

The final rule requires that no straight baseline should be drawn in a manner which cuts off the territorial sea of another country from the high seas or an exclusive economic zone. This is the only unambiguous rule.

COMMON BREACHES OF ARTICLE 7

There are five common breaches of the rules set out in Article 7. First several states including Albania, Australia, Burma, Colombia, Cuba, Iceland, Ireland, Italy, Mauritania, Senegal and Sweden have drawn all or sections of their straight baselines on coasts which are smooth rather than deeply indented. Often such infringements do not push the outer limit of the territorial seas away from the coast to any

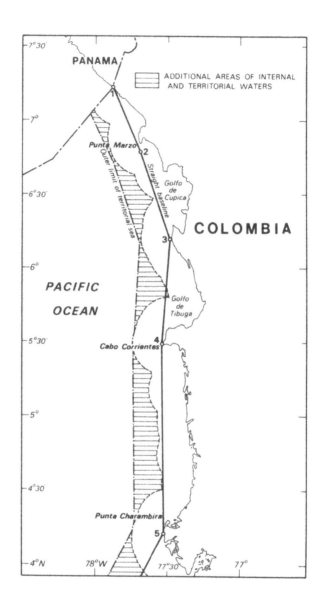

Figure 3.1. A section of Colombia's straight baselines on its Pacific coast

great extent but they do have the effect of weakening the force of this provision. Figure 3.1 shows part of the Colombian straight baseline in the Pacific Ocean which was declared on 13 June 1984. (9) The coast between the boundary with Panama and Punta Charambira consists of two main parts separated by Cabo Corrientes. To the north the coast is cliffed while to the south the narrow coastal plain is fringed with mangroves. (10) A short section of coral reefs links the two sections in the vicinity of Cabo Corrientes. This coastline possesses a number of gentle embayments such as Golfo de Cupica and the recess bounded by Cabo Corrientes and Punta Charambira but neither of these could be considered to be deep indentations. Further the outer limit of territorial waters generated from the normal baseline does not have a complex configuration.

The second common breach of Article 7 occurs when straight baselines are drawn to connect islands which could not be considered to fringe the coast. Generally the islands in such cases are too few in number or are too distant from the coast to possess the characteristics of a fringe in the immediate vicinity of the coast. Such improper straight baselines have been drawn along parts or all of the coast of Colombia, Ecuador, France, Guinea, Iceland, Iran, Italy, Malta and Vietnam. Figure 3.2 shows part of the straight baseline system in southwest Iceland. These segments were proclaimed on 11 March 1961; the revision of Iceland's straight baselines published on 14 July 1972 did not alter this section. (11) This section of baseline extends from Lundadrangur, which is a rock pinnacle 7 cables off the headland near Dyrholaey, to Gierfuglasker 36 nautical miles to the west, and then passes directly to Eldeyjardrangur 70 nautical miles to the northwest. This last feature is a large rock 8 meters high lying 9 nautical miles off the headland at Reykjanes. (12) This section of baseline surrounds the smoothest section of Iceland's coasts and the regularity of the shore results from the outwash plains and strand flats. (13) Gierfuglasker is one of about sixteen islands and rocks which lie on the Eydjabanki which is a linear extension of the coastal bank. Eldeyjardrangur is one of about 10 rocks and islands which extend southwest of Reykjanes. While it might be argued that these two linear groups of islands fringe the short sections of coast which lie closest to them it is quite unreasonable to assert that they fringe the entire

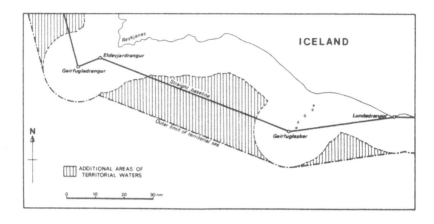

Figure 3.2 A section of Iceland's straight baselines on its southern coast.

coastline of more than 70 nautical miles which lies between them.

The third breach in Article 7 which is becoming increasingly common is the location of basepoints in the sea rather than on or above the low-water mark. There are three situations in which this unjustified action has been taken. First some neighbouring countries have created common origins for the straight baselines. For example, the northern terminus of the Finnish straight baseline proclaimed on 18 August 1956 is located on a straight line joining a Finnish and a Swedish islet at the head of the Gulf of Bothnia. The exact position of this common basepoint occurs at the intersection of the median line between the two countries and the line between the two islets. Similar arrangements have been made by Sweden and Norway and Denmark and West Germany. (14) Second some countries have located the terminus of a straight baseline in the manner just described without apparently consulting their neighbours. For example Chile has terminated its straight baseline in the south at a point in the Beagle Channel fixed by the Court of Arbitration which adjudicated the dispute over islands between Chile and Argentina. (15) On 21 July 1973 Iran proclaimed straight baselines in the Persian/Arabian Gulf and the Gulf of Oman. The western terminus is fixed at the median point in the mouth of the Shatt al Arab, which forms the disputed boundary with Iraq, and the eastern terminus is fixed east of the median point on a closing line of Gwatar Bay which is shared with Pakistan. Third some countries have fixed turning points or termini of straight baselines in the sea in areas which do not involve adjacent states. In April 1974 Bangladesh proclaimed a straight baseline measuring 221 nautical miles. It is defined by eight points and they are all located in the sea so that the boundary follows closely the course of the 10 fathom isobath except off Cox's Bazar where this isobath moves sharply northwards into the embayment at the eastern end of the delta of the Ganges River. This remarkable proclamation was justified on the ground that this is a very shallow area where the morphology of the seabed is constantly changing and where navigation outside the dredged channels for Chalna and Chittagong is dangerous. India and Burma have rejected the validity of this straight baseline. (16) On 19 May 1978 Guinea Bissau published the third version of its straight baseline. Whereas the first two had very properly hugged the fringing islands forming the Arquipelago dos Bijagos the new version proceeds from Cabo Roxo on the Senegalese boundary to the mouth of the Rio Cajet via two basepoints located on parallel 10°40' north. These points were used by France and Portugal on 12 May 1886 in a treaty which allocated territory in this region. It is outrageous to suggest that this line drawn with the intention of avoiding the need to name each island in the archipelago was intended to be a maritime boundary which was not envisaged until much later in the 20th century.

The fourth breach in Article 7 is caused by continental states surrounding offshore archipelagos by straight baselines. These archipelagos such as Islas Galapagos belonging to Ecuador and the Andaman and Nicobar Islands belonging to India cannot be surrounded by archipelagic baselines because Ecuador and India are not archipelagic baselines states. Therefore any baselines drawn about such offshore groups must be justified in terms of Article 7. Plainly the only method is to argue that one comparatively large island is fringed by all the other islands. This argument seems to work for the Furneaux Group in Bass Strait. However it does not work for Houtman Abrolhos off Western Australia, the Islas Canarias or the Islas Galapagos.

The final serious breach in the rules provided by Article 7 relates to the tendency of some countries to draw the outer limit of the territorial waters from imaginary straight baselines which are never published. Such imaginary baselines can be identified easily because the claims to territorial waters as shown on charts are wider that 12 nautical miles and the outer boundary of this zone consists of straight lines. So far Haiti, North Korea and Malaysia have used this highly irregular technique. (17) Haiti published a map at a scale of 1:5 millions on 6 April 1972 showing the outer edge of its territorial sea defined by ten straight lines joining points for which specific co-ordinates were provided. When parallel lines are traced 12 nautical miles landwards of these ten straight lines it is evident that a straight baseline system corresponds only approximately to the coastline of Haiti. (18)

ARCHIPELAGIC BASELINES

In the 1982 Convention Article 46 provides a definition of archipelagic states and the following article describes the rules which govern the construction of achipelagic baselines.

Archipelagic states are those which consist wholly of one or more archipelagos. Archipelagos are defined as a group of islands including parts of islands, interconnecting waters and other natural features, which are so closely interrelated that they either form an intrinsic geographical, economic and political entity or have been regarded historically as such an entity. The requirement for the islands and waters to be closely interrelated is a matter for subjective judgement and there are 35 archipelagic states which could be considered to meet the definition of Article 46.

Article 47 has nine sections which set out five tests which archipelagic baselines must satisfy, specify safeguards for adjacent states which might be disadvantaged by such baselines, and stipulate how such baselines should be publicised.

Three of the five tests are incapable of consistent objective interpretation. First the main islands must be enclosed, second not more that three percent of the baseline legs may exceed 100 nautical miles in length and third the baselines must not depart from the general configuration of the archipelago to any considerable extent. The word 'main' could apply to the largest islands, the most populous islands, the islands which are most economically productive or the islands which are pre-eminent in an historical or cultural sense. The rule that only three per cent of baseline legs may exceed 100 nautical miles in length has a superficial exactness. However if a state needs five legs of that length it only has to ensure that it creates at least 167-200 legs by using more short legs than are strictly necessary to preserve the general configuration of the archipelago.

One test requires that the baselines must enclose an area of water which is at least as large as the area of land enclosed but not more than nine times that land area. This test should be capable of consistent objective application but its interpretation has been qualified. For the purpose of this calculation water enclosed or nearly enclosed by a chain of limestone islands and drying reefs may be counted as land. It is believed that the second variation was introduced for the sole benefit of The Bahamas.

The fifth test prohibits any leg of an archipelagic baseline from exceeding 125 nautical miles in length. Basepoints may be fixed on low-tide elevations lying with the territorial seas generated from land and all other low-tide elevations which are surmounted by a lighthouse or similar installation. (19)

Fifteen archipelagic states are prevented from drawing archipelagic baselines because they cannot enclose an area of water equal to the area of land. These countries are Australia, Cuba, Haiti, Iceland, Ireland, Japan, Madagascar, Malta, New Zealand, Singapore, Sri Lanka, Taiwan, Trinidad and Tobago, the United Kingdom and Western Samoa. Eight of those countries have drawn straight baselines along all or part of their coasts. Cuba, Haiti and Malta have drawn straight baselines around their entire coast even though in some sections they are not justified.

Tuvalu, Mauritius and Kiribati cannot draw archipelagic baselines because the area of water which such lines would enclose would be more than nine times the area of land.

Fiji, Papua New Guinea, Seychelles, Solomon Islands and Tonga could declare archipelagic baselines around some of their islands and Fiji, Papua New Guinea and the Solomon Islands have already taken this action.

Twelve countries can surround their entire territory with archipelagic baselines. They are Antigua, The Bahamas, Cape Verde, Comoros, Grenada, Indonesia, Jamaica, Maldives, Philippines, Sao Tome, and Principe and Vanuatu have already drawn these baselines. If Vanuatu ever succeeded in acquiring Hunter and Matthew Islands from New Caledonia they could not be incorporated into the existing archipelagic baseline system.

CONCLUSIONS

While Article 7 of the 1982 Convention is now honoured in the breach as much as in the observance, breaches of Article 47 by archipelagic states have been so minor as to be quite unimportant. This situation may not continue because it is known that some archipelagic states, such as Kiribati, regard the formulae and measurements in Article 47 as abitrary and without binding force for those archipelagic states which were too poor to involve themselves in the negotiations which produced those rules. They are considered to be rules for the powerful archipelagos such as Indonesia and the Philippines.

While no country has yet drawn straight baselines around highly unstable deltaic coasts this justification might be used in the future. If it does then it is possible that some countries might try to make the application retrospective so that they can offset serious erosion which occurred during say the past 50 years. Other countries might also try to apply the rule to coasts which lack deltas but which are nevertheless highly unstable. Volcanic coasts and tundra coasts may be targets for such variations in the rules.

The chief effect of drawing long improper legs of straight baselines is to augment the areas which fall within internal and territorial waters. Only when the leg is very long are significant areas of exclusive economic zone or potential continental margin gained. The actual gain in any case will depend on the length of the baseline leg, its distance from the coast, the configuration of the coast in that area and the nearness of conflicting claims from neighbouring states. No formula can take all these elements into account satisfactorily. However if attention is focussed on maritime gains seawards of the straight baseline's leg which is at least 24 nautical miles from the coast of islands or the mainland it is possible to predict gains for various zones according to the length of the leg. (20) This information is provided in Table 3.1. The zone which is 350 nautical miles in width is included because that measurement provides one of the absolute limits of claims to the continental margin.

Table 3.1 <u>Seaward areas in square nautical miles gained by unimpeded national claims from a single leg of known length in a straight baseline</u>

Length of leg of a straight baseline in nautical miles

		24	50	100	125	150	200
Width of claimed zone in nautical miles	12	61	373	973	1273	1573	2173
	200	3	27	209	406	703	1735
	350	1	17	113	232	401	971

The table shows that when the legs are short the major increase occurs in the area of territorial waters. As the legs become longer the rate at which the wider zones augment national claims exceeds the rate at which territorial seas are increased. According to the 1958 Convention any increase in the area of territorial waters and internal waters diminishes the area of high seas. Any reduction in the high seas according to the 1982 Convention can only occur by increasing the exclusive economic zone. If straight baselines permit wider claims to the continental margin then it is The Area which is to be subject to international control which is reduced.

The improper use of straight baselines has not so far provoked any serious international disagreement or conflict; the American-Libyan dog-fight some years ago was over an improper closing line. This situation will continue, providing the claimant states do not enforce regulations strictly in the areas gained by using the straight baselines. However, the fact that Article 7 is now effectively a dead-letter may mean that it will be harder to defend the strict interpretation of other articles in the 1982 Convention.

NOTES

1. Case concerning the Continental Shelf (Libyan Arab Jamahiriya/Malta), Judgement of 3 June 1985, (International Court of Justice, The Hague, 1985), 48.

2. J.R.V. Prescott, The Maritime Political Boundaries of the World (Methuen, London, 1985), 298 and 313.

3. The Law of the Sea: Official text of the United Nations Convention on the Law of the Sea with annexes and index (United Nations, New York, 1983), 4.

4. Reports of Judgements, Advisory Opinions and Orders, (International Court of Justice, The Hague, 1951), 116-206.

5. See note 3.

6. Reports of Judgements, Advisory Opinions and Orders, (International Court of Justice, The Hague, 1951), 142.

7. R.D. Hodgson, and L.M. Alexander, Towards an Objective Analysis of Special Circumstances, (Law of the Sea Institute, Occasional Paper no.13, Rhode Island, 1972), 37.

8. Reports of Judgements, Advisory Opinions and Orders, (International Court of Justice, The Hague, 1951), 133.

9. 'Straight baselines: Colombia' Limits in the Seas no.103, (The Geographer, Department of State, Washington DC, 1985).

10. E.C.F. Bird, and M.L. Schwartz, The World's Coastline, (Van Nostrand Reinholt Company, New York, 1985), 45-7.

11. 'Straight baselines: Iceland', Limits in the Seas no.34 (The Geographer, Department of State: Washington DC, 1971).

'Straight baselines: Iceland' Limits in the Seas no.34 revised (The Geographer, Department of State: Washington DC, 1974).

12. Arctic Pilot volume II, (The Hydrographer of the Navy, Taunton, 1975) 51-7.

13. E.C.F. Bird, and M.L. Schwartz The World's Coastline (Van Nostrand Reinholt Company, New York, 1985) 267-8.

14. J.R.V. Prescott, The Maritime Political
Boundaries of the World, (Methuen, London, 1985),
278.

15. Ibid, 203.

16. Ibid, 163 and 166.

17. J.R.V. Prescott, 'Maritime jurisdictional
issues' in George Kent and Mark Valencia (eds),
Marine Policy in Southeast Asia (University of
California Press, Los Angeles, 1985), 58-97.

Maritime boundaries in the North Pacific Region,
Research memorandum No.1/410 (83), (Office of
National Assessments, Canberra, 1983).

18. 'Straight baselines: Haiti' Limits in the Seas
no.51 (The Geographer, Department of State:
Washington DC , 1973).

19. J.R.V. Prescott, The Maritime Political
Boundaries of the World, (Methuen, London, 1985),
70-2.

20. I am grateful to John Barton and Sidney
Clifton, formerly of the University of Melbourne for
calculating the formula on which Table 3.1 is based.

Chapter 4

THE IMPORTANCE OF GEOGRAPHICAL SCALE IN CONSIDERING
OFFSHORE BOUNDARY PROBLEMS

Ewan Anderson

The United Nations Convention on the Law of the
Sea (1982) was formulated for application at a
global scale, but its implementation depends very
much upon local geography. This leads to many
problems of scale when actual examples are
considered. It is realised that such a problem is
merely one aspect of the general/specific dichotomy
in scientific analysis. However, it is contended
that in the field of international boundary delimitation 'ground truth' is vital, and in the quest
for generally applicable laws, is often overlooked.
    In recognising that every case is monotypic,
the International Court has become more pragmatic in
its workings. The key principles are those which
produce the most equitable result. Ideas such as
proportionality and equality are viewed as desirable
factors rather than principles. Furthermore, it has
become increasingly clear that geography has a key
role to play, particularly through geomorphology and
geomorphometry. At present for the requirements of
law, both subjects tend to lack sophistication.
However, developments in surveying and data handling
promise far greater precision in land-form
description, both qualitative and quantitative, in
the future.
    Geomorphology in particular has played a
significant role in several recent cases, and
examples related to problems of scale are discussed.

'OPPOSITE' COASTLINES

    Several basic questions must be posed about the
term 'opposite'. To be opposite, must coastlines be
of approximately the same length? Should a line
drawn normal to one be normal to the other? What is

52

the difference between opposite and parallel?
    In the case of Malta and Libya (International
Court of Justice, 1985), two totally different
coastlines were being matched. The coastline of the
Maltese Islands is extremely limited in extent,
angular and sharply defined. That of Libya is the
longest of the North African coast and is in general
depositional and gently curved. Furthermore, the
length of the Libyan coastline which could be in any
way related, is approximately twenty times that of
the Maltese coast. The discrepancy in coastline
length is indeed so great that it is impossible to
make comparisons on the same chart. However, using
the coastal length ratio of 1:20, straight baselines with a minimum length of 0.5 nautical miles
(nml) may be compared with similar baselines of
10 nml along the Libyan coast. The limiting lengths
and approximate bearings of the relevant coastlines
of Malta which could be considered to be opposite
Libya are shown on Figure 4.1.
    To allow meaningful measurement to establish
whether coastlines are opposite, it is obviously
necessary to subdivide the coast into straight
lengths. Using natural facets down to a length of
one nml and land configuration to provide a guide
for bay enclosure, the trend of the coastline can be
ascertained. For the Maltese coastline, each facet
is measured to the nearest 0.1 nml and the bearing
of the line normal to it to the nearest 0.5° (Figure
4.2). Arrowheads on the normals indicate that a
bearing line produced would strike the coastline of
Libya. Thus on this scale a total coastal baseline
length of 17.1 nml could be deemed to face Libya.
    Again using structural trends, but in this case
the longest possible baselines, the total length
facing Libya is found to be 15.7 nml (Figure 4.3).
Obviously, since the discrepancy between the
coastlines is so great, it seems reasonable to take
the longest possible facets.
    Relevant coastlines on a larger scale, sub-
divided into micro-facets down to a scale of 0.5 nml
are shown in Figures 4.4, 4.5 and 4.6. Arrowheads on
the normals indicate those bearings which produced
would strike the coastline of Libya. A half arrow-
head indicates that this applies to only half of the
facet. On this scale, 14.5 nml of coastal baseline
could be deemed to face Libya.
    Thus, it can be readily seen that whether or
not a coastline is opposite depends very much upon
the scale of the geometry used for baseline
facet production. It must be considered that any

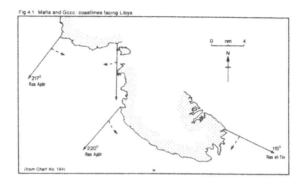

Figure 4.1. The maximum extent of the coastlines of Malta and Gozo facing Libya

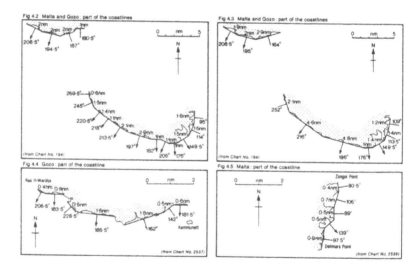

Figures 4.2 to 4.5. Coastline sections opposite the coast of Libya, using baselines of different lengths

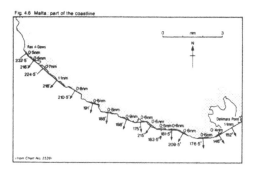

Figure 4.6. Coastline sections opposite the coast of Libya using the shortest baselines

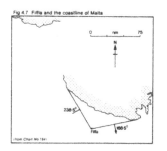

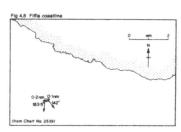

Figures 4.7 and 4.8. Baselines opposite the Libyan coast, using Filfla island

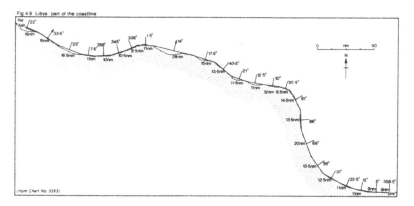

Figure 4.9. Baselines opposite the coastlines of Malta and Gozo

such facet shorter than 1.5 nml can only be regarded
as a micro-feature and cannot be thought to indicate
a coastal trend. To illustrate the point further, if
the coastal baseline is drawn to include Filfla
Island, the length of coastline of the island of
Malta alone which faces Libya is reduced from 10.6
to 8.2 nml. Figure 4.7. It is interesting to move to
an even smaller scale and consider the island of
Filfla alone. On the largest scale chart available,
it is just possible to discern the trend of coastal
baselines on Filfa and these can be seen to face
long stretches of the Libyan coastline, Figure 4.8.

The Libyan coastline cannot be analysed on the
same scale as no suitable charts are available.
Furthermore, changes in trend, with one or two
exceptions, are so slight that it is possible to
construct far longer baseline facets. Since it is
also predominantly depositional, the Libyan coastline is far less precisely demarcated, because along
many stretches the boundary between land and sea is
not immediately obvious.

The only length of coastline which could not be
considered in any way opposite Malta is that from
Ras Ashdir to Ras et-Tin from which point the
orientation is predominantly easterly. Although much
of this is gently curved, by using facets of 10 nml
as justified it is possible to make a meaningful
subdivision. It can be seen (Figure 4.9) that only
two facets with a total length of 44 nml can be
considered as facing Malta. Furthermore, while the
westerly of the two facets represents a clear and
obvious coastal trend, the other depends completely
on whether the bay is enclosed or whether it is
deemed to subsume two separate facets.

By producing bearings normal to facets, it is
possible to illustrate the vast discrepancy in
coastlines which might be reckoned to be opposite.
When all the bearings normal to facets of the
Maltese coastline, identified as facing Libya, are
produced, (Figure 4.10) and normals constructed at
the Libyan end, it is revealed that only two facets
are indeed opposite. However, in this context, the
term opposite requires further definition. While
parts of the facets may well be opposite, in normal
parlance most must be regarded as 'parallel'. This
point is illustrated in Figure 4.11. Bearings from
Malta of 213.5° and 194° are parallel and in part
opposite. These two bearings link lengths of 4.1 nml
along the Maltese coastline, with a total facet
length of approximately 73 nml on the Libyan coast.
Allowing a greater discrepancy between the actual

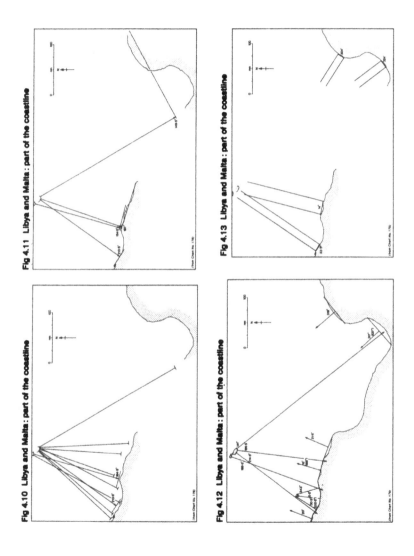

Fig 4.10 Libya and Malta : part of the coastline

Fig 4.11 Libya and Malta : part of the coastline

Fig 4.12 Libya and Malta : part of the coastline

Fig 4.13 Libya and Malta : part of the coastline

angles of opposite coastlines, bearings from Malta of 197° and even possibly 149.5° can be seen to provide a coarse fit. At another scale (Figure 4.12) the coastline of the Maltese Islands can be reduced to four relevant facets. The production of normals from these indicates that rather more of the Libyan coastline is parallel, provided an appropriate subdivision of the eastern Libyan coastline is produced.

In Figure 4.13 the projection of the opposite and parallel facets illustrates the vast discrepancy in scale. For example, the facet, the normal from which bears 14° from Libya, would, when projected, subsume all the relevant coastline of Malta, together with part of that of Gozo.

ORIENTATION

If natural prolongation of coastline is considered important in legal disputes, then the orientation of that coastline must be assessed. In any other than a straight coastline, this presents obvious problems. As with the previous analysis, if the coastal irregularities are smoothed into straight coastline facets, it is possible to make judgements, but these will depend very much upon the draughtsmanship involved. In the case of Tunisia and Libya, two adjacent states, the orientation of the coastlines was considered important.

If the coastlines of both Libya and Tunisia are smoothed to one straight line it can be seen that the Tunisian coast faces east-north-east and the Libyan coast a few degrees east of north (Figure 4.14). If the smoothing takes account of natural features such as bays and headlands and a limited number of straight lines is produced (Figure 4.15), dominant trends can be discerned. In the case of the Tunisian coastline, this is north-eastern and south-eastern, with only the Gulf of Hammamet facing due east. The Libyan coastline largely faces due north, except for the eastern sector, which is angled slightly towards the east.

Perhaps a more objective method is to subdivide the coastline regularly and aggregate the orientations of each short length (Figure 4.16). The resulting orientation diagrams indicate that the Libyan coast is predominantly north-facing and almost exclusively orientated between north-north-west and north-north-east. In the case of the Tunisian coastline, there is a wide spread of

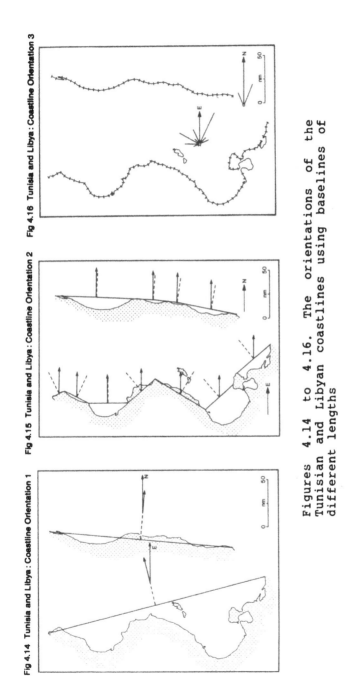

Fig 4.14 Tunisia and Libya : Coastline Orientation 1

Fig 4.15 Tunisia and Libya : Coastline Orientation 2

Fig 4.16 Tunisia and Libya : Coastline Orientation 3

Figures 4.14 to 4.16. The orientations of the Tunisian and Libyan coastlines using baselines of different lengths predominant orientation, with a distinct bias towards the north of east.

Thus, it is clear again that in assessing coastline orientation, problems of scale are paramount. While mathematical methods of averaging can be employed, it is considered vital that the natural features are given due weight. While geography may be neutral, it should not be refashioned.

ADJACENT COASTLINES

The demarcation of offshore boundaries between adjacent states would seem to be a rather simpler task than that between opposite states. However, of great importance must be the angle of projection of the boundary from the coastline. It might be thought appropriate that this should follow the trend of the land boundary, but again scale factors would enter the argument. On a small scale chart, the boundary between Libya and Tunisia would appear to trend north-south. On a larger chart, it can be seen that for the last part of its length, the boundary is inclined eastwards. Since this particular difference reflected the area of sea in contention it was an important consideration.

SCALE AND EQUITY

While it may be agreed that equitable principles and equity are not the same thing, it has been considered that large differences in areal extent or length of coastline between countries should play a part in offshore apportionment. Furthermore, it must be remembered that since the smaller the state, the longer proportionately is its boundary, micro-states already possess an innate advantage. Thus, in the case of islands such as Bahrain and Malta, by the provision of at least some median boundaries, they obtain jurisdiction over areas of water far larger than their own land area. The coastlines of Libya and Malta have already been discussed in this context, and the two states are obviously on a totally different scale.

The Ionian Sea is the one large water body in the Mediterranean uninterrupted by islands. It is therefore possible to contemplate theoretically a comparatively uncomplicated subdivision. If the Ionian Sea is envisaged as basically triangular then the lengths of Maltese coastline contiguous with one side would be tiny compared with those of Italy, Greece and Libya. Similarly, if a solution such as that which obtains in Antarctica is employed, with a sectoral division the rather modest amount of sea accruing to Malta as opposed to Libya is clear. It is obvious that unless scale is at least to an extent taken into account, then such solutions are likely to be inequitable.

ENCLOSURES AND ALLOWANCES

Scale is also a vital factor in considering enclosures of various features and allowances for them (1). Since 1958 a bay has been defined and distinguished from a mere indentation, but this still does not solve problems of orientation associated with the direction of the closing line. There are also historic bays and those shared by more than one littoral state for which no judgements have yet been made. Therefore how and whether a legal bay can be enclosed is reasonably well defined, but in the final analysis, local geographical factors are still vital. For example, the enclosure of the Gulf of Sirte has been a contentious issue since 1973, but in some ways even less clear, is exactly where the closing line might be drawn. Is there a natural line of enclosure? Similarly, the position of enclosures across river mouths will depend very much upon the interpretation of the local geomorphology.

Many other features might be mentioned, but one of particular relevance concerns low tide elevations and islands. There has been debate about what actually constitutes both and the importance with regard to sea bed jurisdiction is obvious. Again, the scale of the chart used and of the feature itself may well be a crucial determinant.

Mention should also be made of man-made structures, since harbour constructions and other attachments to the coast may be used for baseline construction. Thus, a feature too small to appear on many charts may be a component in boundary making. One good example of this occurred in the Libyan-Tunisia case, in which the Greco Bank was located almost centrally in the area of contention. This small feature, from many angles hardly discernable above the shelf had long been important for sponge fishing and a series of significant historic precedents had been set. Thus, the location of this particular feature provided important evidence in the case.

OTHER ASPECTS OF SCALE

The significance of scale can be demonstrated clearly in many other respects. Distortions on various scales can of course be introduced by the differing use of map projections. The position of a line of judgement of the International Court such as that between Libya and Tunisia will appear very different, according to the particular projection used. This is obviously crucial for establishing its exact location on the surface of the earth.

On a totally different level, there is a growing tendency for problems to be examined academically at least on a regional rather than a global scale (2). Since there are no natural boundaries at sea, many offshore developments, ranging from mining to pollution, affect a number of neighbouring countries. In such cases the application of the Law of the Sea principles can be seen as scale-specific.

CONCLUSIONS

It has been suggested that in offshore boundary considerations, the local geographical environment is vital. Boundaries must be related to geomophological features and the scale of variation in these which is considered significant must be taken into account. Since boundary lines may well be crucial for mining or oil extraction, their exact position must be known. However, the effects of such exploitation may well be specific and regional while legislation needs to be general and global. Nevertheless, to produce equitable global legislation, local variables must be taken into account and once that legislation has been enacted, it is at the local scale that its effects will be felt.

NOTES

1. R.R. Churchill and A.V. Lowe, The Law of the Sea, (Manchester University Press, 1983), 24-44.

2. B.A. Boczek, 'Global and regional approaches to the protection and preservation of the marine environment', Case Western Reserve Journal of International Law, 16, 1, (1984), 39-70.

Chapter 5

THE LIMITS OF THE AREA BEYOND NATIONAL JURISDICTION - SOME PROBLEMS WITH PARTICULAR REFERENCE TO THE ROLE OF THE COMMISSION ON THE LIMITS OF THE CONTINENTAL SHELF

Piers R.R. Gardiner

INTRODUCTION

Although the United Nations Convention on the Law of the Sea, 1982 (1) has yet to enter into force, it has provided for the first time a resource-related comprehensive demarcation of the seabed and subsoil into two major zones viz the areas within and beyond the limits of national jurisdiction. As pointed out in the Preamble to the UN Convention, the latter area, with its mineral wealth potential has been recognised as the common heritage of mankind to be exploited for the benefit of mankind as a whole. For the international community in general, therefore, the boundary of the area beyond national jurisdiction (the Area) is of major significance. Not only should this boundary be accurately located, but it should also be internationally agreed, so that the full extent of the Area is preserved and subsequent 'creeping' jurisdiction prevented. It is recognised, however, that the concept of the Area, which has resulted from the new doctrine of the 'Common Heritage of Mankind', is still the subject of contoversy. (2) International customary law has yet to reflect the full impact of the UN Convention, but it would seem likely that future differences will be concerned with specific legal regimes in the various maritime zones recognised in the UN Convention rather than the boundaries between such zones. Much has already been written about the delimitation of the areas of national jurisdiction during the preparatory work that led to the UN Convention and subsequently, but this has tended to be from the perspective of coastal states and influenced by their individual problems. The purpose of this chapter is therefore to briefly focus on the relevant provisions in the UN Convention concerning the limits of the area beyond national jurisdiction from the viewpoint of the international community, evaluate possible legal problems arising from the text of the provisions as regards the location of such limits, and assess the way in which the interests of the international community are likely to be safeguarded under the terms of the UN Convention. The situation concerning delimitation of maritime boundaries between states has been discussed elsewhere and is not considered here. (3)

THE LIMITS OF THE AREA BEYOND NATIONAL JURISDICTION

The boundary of the Area will, under the terms of the UN Convention, be a linear amalgam of two different limits. These are firstly the outer limit of the Exclusive Economic Zone (EEZ), and secondly the outer limit of the Continental Shelf where it is deemed to extend seaward beyond the limits of the EEZ. In both cases the limits will be proposed by coastal states under specific rules as part of their determination of their limits of national juris-diction, resulting in the intriguing situation that the geographical extent of the Area for the inter-national community will effectively be determined on a residual basis, although there is scope for reactive assessment as indicated below.

THE OUTER LIMIT OF THE EEZ

Under Article 57 of the Convention, it is stated that the EEZ 'shall not extend beyond 200 nautical miles from the baselines from which the breadth of the territorial sea is measured'. There are two elements here which warrant consideration, namely, the manner in which the baselines are determined and drawn, and the method of location of the outer limit. Both will affect the precise location of the perimeter of the area.

The rules for drawing baselines for coastal states are given in Articles 7, 9 and 10 of the Convention, and under Article 47 for archipelagic states, and although to date the rules for archipelagic states have essentially been followed, it is also likely that future breaches will occur in this regard. The effect of these divergences from the method of location of baselines as laid down in the UN Convention is to encroach on the Area. The amount of reduction of the Area, if permitted, has been shown by Prescott to be related to the length of the straight baseline leg, becoming significant for any leg longer than 24 nautical miles. A number of countries, including Bangladesh, Iceland and Colombia have declared baselines substantially in excess of this figure. (4)

The method of location of the outer limit of the EEZ was a matter for concern at the conference negotiations (UNCLOS III), since differences in the line positioning could arise if states used different datums and spheroids for representational purposes. (5) Under Article 75 of the Convention the situation has been clarified to the extent that States are required to suitably publicise their charts or point locations, and if the latter, to specify the geodetic datum. Unfortunately problems could still exist, since the required nature of the lines between points is not stated, and it would appear from the wording of the article that if only charts are supplied by the coastal state no agreed cartographic method need be demonstrated. As a result, there could be errors at the expense of the boundary of the Area.

THE OUTER LIMIT OF THE CONTINENTAL SHELF

The manner in which this boundary is to be delimited is given in Article 76 of the UN Convention (see Appendix). This article dealt with one of the fundamental issues at UNCLOS III, and its complex nature is a reflection of the diverse facets of this technical topic and the negotiated compromises (6) that led to the present text. Several analyses of the application of Article 76 have recently been provided, (7, 8, 9) and while there would seem little difficulty with technical aspects the language of some of the provisions is open to more than one interpretation. Since these have a direct bearing on the location of the outer limit and hence the extent of the Area, the possible effects of alternative interpretations merit further discussion.

In paragraph 1, the juridical continental shelf is defined as either relating to the physical entity of the continental margin, or 'extending to a distance of 200 nautical miles from the baselines from which the breadth of the territorial sea is measured'. A situation, however, could arise whereby the application of the 200 nautical mile limit by one coastal state could result in the overlap onto a physically separate continental margin (Figure 5.1). By arguing that its juridical continental shelf now had a submerged prolongation, the coastal state could lay claim to additional areas that would be delimited by the specific methods given in Article 76 unless there was a closer boundary of another coastal state. Such an approach is, however, contrary to the reasoning behind the provisions of article 76, (6) as is implied by paragraph 3, which was intended to ensure that the legal continental shelf of a coastal state could only extend beyond 200 nautical miles with reference to its <u>own</u> continental margin. The additional area could be substantial, and if such an approach were successful would detract from what would seem reasonably to be part of the Area.

In paragraph 4, the coastal state is provided with two methods by which it can accurately establish the limits of its national jurisdiction. As reasonably interpreted, and intended, (6, 8) the coastal state can have the option of using either method for any part of its continental margin to locate a fixed point, and can change from one method to the next for adjacent fixed points. Alternatively, although unlikely, the text could be taken as indicating that the coastal state could only use one method throughout in determining the outer edge of the continental margin.

In paragraph 5, two options (depth or distance) are given for limits beyond which national jurisdiction over the continental shelf cannot extend (subject to the provisions of paragraph 6). It is, however, not entirely clear if the coastal state will be restricted to consistently applying one option, or to whichever boundary is closer inshore, or be free to select whichever limit extends further seaward at any particular point. The differences in interpretation could have a significant effect on the extent of the Area, since there need be no relationship between depth and distance. There seems little doubt, however, that the latter interpretation is implied in the wording of the text, since the second option was apparently included to cover situations where there was a shallow wide physical continental shelf (sensu stricto). These would be unreasonably truncated short of the foot of the physical continental slope if the distance limit alone were to apply, a view which is indirectly supported by the last sentence of paragraph 6. Naturally, however, if there was a complete physical

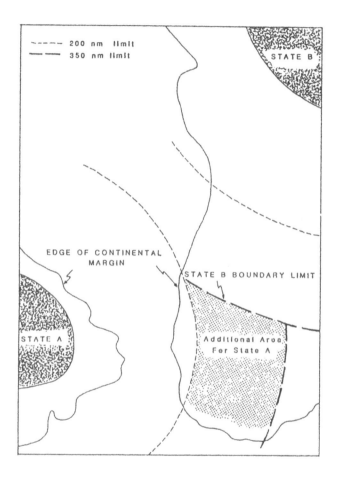

Figure 5.1. Simplified diagram illustrating possible additional area of national jurisdiction that could be claimed by a coastal state beyond its own continental margin. For clarity the 350 nautical mile limit is taken as the outermost boundary, and boundary limits within the physical continental margin as would be determined under paragraph 4 of article 76 are ignored.

break in a shallow continental margin, such as a deep trough, then the limit of national jurisdiction would be terminated at this point.

Paragraph 6 may also pose difficulties, due in part to its introduction at a late stage in the UNCLOS III negotiations as compromise language, and has been analysed in detail by McKelvey. (8) The problem revolves around the identification of submarine elevations to which the 350 nautical mile limit would apply. It could be argued that submarine ridges that are part of the ocean floor are implicitly recognised as such in this paragraph, and where they are the natural prolongation of a land mass or island such as Iceland, could fall under national jurisdiction to a 350 nautical mile limit. However, as McKelvey (8) has pointed out, this interpretation would be in conflict with paragraph 3, which specifically excludes the ocean floor with its oceanic ridges, and hence he considers that it would be difficult to sustain. Since the interpretation would in these cases mean the difference between a national outer limit of 200 nautical miles or 350 nautical miles along a submarine ridge, it is clearly of importance as regards the limits of the Area.

Of the remaining paragraphs, the accurate location of the boundary lines, as specified in paragraphs 7 and 9, (see also Article 84) would be subject to the same concerns as have been outlined earlier with regard to the outer limit of the EEZ.

VERIFICATION ASPECTS AND THE ROLE OF THE COMMISSION ON THE LIMITS OF THE CONTINENTAL SHELF

As indicated earlier, there are a a number of possible situations that could result in coastal states delineating boundaries that may encroach on the geographical extent of the Area. There would therefore seem to be a clear need for a system which will reassure the international community that the limits of the Area are properly and accurately located. Although the International Sea-Bed Authority (the Authority) will be established under the UN Convention to organise and control activities in the Area (Article 157), it has no role as regards boundary demarcation. Provision, however, exists under the UN Convention for appropriate assessment of the outer limit of the Continental Shelf, the need having been identified at an early stage in UNCLOS III, (6) but the situation regarding the outer limit of the EEZ is less clear. These aspects are discussed in turn.

Paragraph 8 of Article 76 requires that proposals concerning the outer limits of the continental shelf be submitted by the coastal state to the Commission on the limits of the Continental Shelf. As set up under Annex II of the UN Convention (see Appendix), the Commission as an independent body has both an advisory and verification role. The latter is a reactive process in that if it does not recommend (in effect approve) the submission, or part of it in the context of Article 76, the coastal state is required to make a revised or new submission for the proposed boundary. However, those limits which are established by the coastal state on the basis of recommendations by the Commission would be final and binding. The process was envisaged by its proponents at UNCLOS III (6) as being a narrowing down 'ping-pong' procedure in cases where the full length of the outer limit was not deemed to be satisfactorily located, whereby the coastal state would bring back revised submissions over progressively shorter sectors until the total boundary limit was recommended by the Commission. Since the Commission would be examining all technical aspects, the possible problems of depiction on charts referred to earlier in the context of Article 84 are likely to be satisfactorily resolved during this process.

As Brown (9) has pointed out, however, a problem could arise if the coastal state finds the recommendations of the Commission unacceptable, since it is not in the mandate of the Commission to impose a boundary line. If the coastal state proceeded with the establishment of its boundary, contrary to the provisions of Article 76(8), this could be disputed by another state, with the obligation of binding settlement in accordance with Part XV (Settlement of Disputes) of the UN Convention. Alternatively, the coastal state submission could remain in limbo, with subsequent difficulties in the recognition of the boundary of the Area.

Once the Commission is established, it would seem necessary for it to consider immediately the interpretive problems in the language of Article 76, and issue appropriate guidelines to coastal states intending to make boundary submissions. In this way the difficulties outlined above can be readily overcome, with consequent reassurance to the international community that the boundary of the Area will not be unreasonably eroded. The Commission should also be aware of areas of the continental shelf beyond 200 nautical miles that are being claimed by more than one coastal state, and consider how it will treat submissions from them concerning the outer limit if these are made before the dispute is resolved. The basis of submissions and different cut-off limits may result in different boundaries being put forward, with subsequent evaluation difficulties for the Commission.

Continental margins beyond 200 nautical miles only comprise some six per cent of the sea-floor, (10) so in relative boundary terms the accurate location of the outer limit of the EEZ is far more important for the Area as a whole. Since there is no provision in the UN Convention for a body with functions similar to the Commission on the limits of the Continental Shelf to consider EEZ limits, it would seem to fall to other states to dispute the charted limits lodged by coastal states under Article 75. The matter would then be resolved under the Settlement of Dispute procedures of the UN Convention. There is, however, a possible alternative. In situations where the continental margin of the coastal state does not extend beyond 200 nautical miles from the baselines, delimitations of the juridicial continental shelf is made on the 200 nautical mile criterion. The boundary is therefore the same as the outer limit of the EEZ. The Commission could consider that its role included the assessment of limits drawn on this basis, if the text of Article 76(8) was interpreted to mean all information on the limits of the continental shelf beyond 200 nautical miles. In this case the information would be that there was no extension, and the 200 nautical mile boundary would be submitted for recommendation. Such an approach would mean that all boundaries to the Area would be submitted to an independent body, which would perhaps prove to be the most reasonable way of satisfying the international community that the boundary of the 'common heritage of mankind' was suitably safeguarded.

CONCLUSIONS

Under the UN Convention the boundaries of the Area may not always be accurately located due to lack of compliance with, or interpretive problems with, the relevant articles. This could result in a reduction of what should be the correct geographical extent of the Area. The Commission on the Limits of the Continental Shelf is seen as having a vital role in ensuring that there is uniformity of application of Article 76 (Definition of the Continental Shelf) and that the boundary to the outer limit of the juridical continental shelf is equitably verified. Possible encroachment into the Area by coastal state delineations of the outer limit of the EEZ can, however, only be disputed by other states, unless the Commission on the Limits of the Continental shelf is able to consider such delineations in the context of the continental shelf regime.

NOTES

1. United Nations Convention of the Law of the Sea, (United Nations, New York, 1983), Sales No.E.83.V.S., 224pp.

2. E.D. Brown, 'The UN Convention on the Law of the Sea 1982', Journal of Energy and Natural Resources Law, 2(4) (1984) 258-282.

3. G.H. Blake, see Chapter 1.

4. J.R.V. Prescott, see chapter 3.

5. R.D. Hodgson, and R.W. Smith, 'The Informal Single Negotiating Text (Committee II): a geographical perspective', Ocean Development and International Law 3(2), (1976) 225-259.

6. P.R.R. Gardiner, 'Reasons and methods for fixing the outer limit of the legal continental shelf beyond 200 nautical miles', Iranian Review of International Relations, 11-12, (1978), 145-170.

7. D.G. Crosby, The UNCLOS III definition of the continental shelf: application to the Canadian offshore, in D.M. Johnston, and N.G. Letalik, (eds) The Law of the Sea and Ocean Industry: New Opportunities and Restraints, Proceedings of the Law of the Sea Institute, (University of Hawaii), (1984), 473-486.

8. V.E. McKelvey, Interpretation of the UNCLOS III definition of the continental shelf, in D.M. Johnson, and N.G. Letalik, (eds), The Law of the Sea and Ocean Industry: New Opportunities and

Restraints, Proceedings of the Law of the Sea Institute, (University of Hawaii, 1984), 465-472.

9. E.D. Brown, The Areas within National Jurisdiction, Sea-bed Energy and Mineral Resources and the Law of the Sea, Volume 1, (Graham and Trotman, London, 1984).

10. A.A. Archer, and P.B. Beazley, 'The geographical implications of the Law of the Sea Conference', Geographical Journal, 141(1), (1975), 1-13.

ACKNOWLEDGEMENTS

A number of aspects presented were refined in disussion with colleagues, particularly Mr R. Keary, whose observations are much appreciated. Grateful thanks are also due to Ms Geraldine Skinner for her wise counsel since the early sessions of UNCLOS III and to her and Dr R.R. Horne for reviewing the manuscript. Any errors or misapprehensions are, however, the responsibility of the author. It is also a pleasure to acknowledge the interest and unfailing patience of Dr Gerald Blake, who kindly permitted me to see the texts of several other contributors, and without whom this contribution would not have been possible. Publication is with the permission of the Director of the Geological Survey of Ireland, but the views expressed are entirely personal and unofficial.

APPENDIX

SELECTED ARTICLES FROM THE UNITED NATIONS CONVENTION OF THE LAW OF THE SEA

Part VI: Continental Shelf

Article 76: Definition of the continental shelf

1. The continental shelf of a coastal State comprises the sea-bed and subsoil of the submarine areas that extend beyond its territorial sea throughout the natural prolongation of its land territory to the outer edge of the continental margin, or to a distance of 200 nautical miles from the baselines from which the breadth of the territorial sea is measured where the outer edge of

the continental margin does not extend up to that distance.

2. The continental shelf of a coastal State shall not extend beyond the limits provided for in paragraphs 4 to 6.

3. The continental margin comprises the submerged prolongation of the land mass of the coastal State, and consists of the sea-bed and subsoil of the shelf, the slope and the rise. It does not include the deep ocean floor with its oceanic ridges or the subsoil thereof.

4. (a) For the purposes of this Convention, the coastal State shall establish the outer edge of the continental margin wherever the margin extends beyond 200 nautical miles from the baselines from which the breadth of the territorial sea is measured, by either:
(i) a line delineated in accordance with paragraph 7 by reference to the outermost fixed points at each of which the thickness of sedimentary rocks is at least 1 per cent of the shortest distance from such point to the foot of the continental slope; or
(ii) a line delineated in accordance with paragraph 7 by reference to fixed points not more than 60 nautical miles from the foot of the continental slope.
(b) In the absence of evidence to the contrary, the foot of the continental slope shall be determined as the point of maximum change in the gradient at its base.

5. The fixed points comprising the line of the outer limits of the continental shelf on the sea-bed, drawn in accordance with paragraph 4 (a)(i) and (ii), either shall not exceed 350 nautical miles from the baselines from which the breadth of the territorial sea is measured or shall not exceed 100 nautical miles from the 2,500 metre isobath, which is a line connecting the depth of 2,500 metres.

6. Notwithstanding the provisions of paragraph 5, on submarine ridges, the outer limit of the continental shelf shall not exceed 350 nautical miles from the baselines from which the breadth of the territorial sea is measured. This paragraph does not apply to submarine elevations that are natural components of the continental margin, such as its plateaux, rises,

caps, banks and spurs.

7. The coastal State shall delineate the outer limits of its continental shelf, where that shelf extends beyond 200 nautical miles from the baselines from which the breadth of the territorial sea is measured, by straight lines not exceeding 60 nautical miles in length, connecting fixed points, defined by coordinates of latitude and longitude.

8. Information on the limits of the continental shelf beyond 200 nautical miles from the baselines from which the breadth of the territorial sea is measured shall be submitted by the coastal State to the Commission on the Limits of the Continental Shelf set up under Annex II on the basis of equitable geographical representation. The Commission shall make recommendations to coastal States on matters related to the establishment of the outer limits of their continental shelf. The limits of the shelf established by a coastal State on the basis of these recomendations shall be final and binding.

9. The coastal State shall deposit with the Secretary-General of the United Nations charts and relevant information, including geodetic data, permanently describing the outer limits of its continental shelf. The Secretary-General shall give due publicity thereto.

10. The provisions of this article are without prejudice to the question of delimitation of the continental shelf between States with opposite or adjacent coasts.

ANNEX II: COMMISSION ON THE LIMITS OF THE CONTINENTAL SHELF

Article 1

In accordance with the provisions of Article 76, a Commission on the Limits of the Continental Shelf beyond 200 nautical miles shall be established in conformity with the following Articles.

Article 3

1. The functions of the Commission shall be:
(a) to consider the data and other material submitted by coastal States concerning the outer

limits of the continental shelf in areas where those limits extend beyond 200 nautical miles, and to make recommendations in accordance with Article 76 and the Statement of Understanding adopted on 29 August 1980 by the Third United Nations Conference on the Law of the Sea.
(b) to provide scientific and technical advice, if requested by the coastal State concerned during the preparation of the data referred to in sub-paragraph (a).

2. The Commission may co-operate, to the extent considered necessary and useful, with the Inter-governmental Oceanographic Commission of UNESCO, the International Hydrographic Organisation and other competent international organisations with a view to exchanging scientific and technical information which might be of assistance in discharging the Commission's responsibilities.

Article 4

Where a coastal State intends to establish, in accordance with article 76, the outer limits of its continental shelf beyond 200 nautical miles, it shall submit particulars of such limits to the Commission along with supporting scientific and technical data as soon as possible but in any case within 10 years of the entry into force of this Convention for that State. The coastal State shall at the same time give the names of any Commission members who have provided it with scientific and technical advice.

Article 5

Unless the Commission decides otherwise, the Commission shall function by way of sub-commissions composed of seven members, appointed in a balanced manner taking into account the specific elements of each submission by a coastal State. Nationals of the coastal State making the submission who are members of the Commission and any Commission member who has assisted a coastal State by providing scientific and technical advice with respect to the delineation shall not be a member of the sub-commission dealing with that submission but has the right to participate as a member in the proceedings of the Commission concerning the said submission. The coastal State which has made a submission to the Commission may send its representatives to participate in the relevant proceedings without the right to vote.

Article 6

1. The sub-commission shall submit its recommendations to the Commission.

2. Approval by the Commission of the recommendations of the sub-commission shall be by a majority of two thirds of Commission members present and voting.

3. The recommendations of the Commission shall be submitted in writing to the coastal State which made the submission and to the Secretary-General of the United Nations.

Article 7

Coastal States shall establish the outer limits of the continental shelf in conformity with the provisions of article 76, paragraph 8, and in accordance with the appropriate national procedures.

Article 8

In the case of disagreement by the coastal State with the recommendations of the Commission, the coastal State shall, within a reasonable time, make a revised or new submission to the Commission.

Article 9

The actions of the Commission shall not prejudice matters relating to delimitation of boundaries between States with opposite or adjacent coasts.

Chapter 6

MARITIME BOUNDARIES AND THE EMERGING REGIONAL BASES
OF WORLD OCEAN MANAGEMENT

Hance D. Smith

INTRODUCTION

The purpose of this chapter is to review the
relationships between the emerging patterns of
maritime boundaries on the one hand, and those of
the regional bases of global ocean management on the
other. It thus looks beyond current pre-occupations
with the technical and legal aspects of boundary delimitation towards the objectives and taxonomy of
boundaries as a tool in the development and management of ocean space and resources. The chapter
begins with consideration of the political and
administrative boundaries which has been a principal
focus of the Third United Nations Conference on the
Law of the Sea (UNCLOS III), followed by discussion
of management for boundaries.

THE DEVELOPMENT OF POLITICAL AND ADMINISTRATIVE
BOUNDARIES

The development of political boundaries has
taken place mainly in relation to the post-World War
II emergence of the Law of the Sea, especially since
the commencement of the UNCLOS III negotiations in
the early 1970s. There are two major boundary types
involved. The most important of these from political
and legal points of view has been the jurisdictional
boundaries established relative to the coastlines of
states. Much of the early history of these is
written in terms of fishing limits. UNCLOS III has
further codified these as a series of boundaries
delimiting both the seabed and waters, the main ones
being those delimiting internal waters, the
territorial sea, continental shelf and exclusive
economic zone. Limited-purpose boundaries relate to

contiguous zones and exclusive fishery zones.

The second group of boundaries are allocation boundaries between states. Only a few of these have been the subject of legal argument, mainly in the North Atlantic, emphasising the point that the greatest allocation pressures exist at the developed core of the global industrial economy. The majority of these boundaries have still to be settled. The main impetus behind the settlements has been offshore oil and gas development rather than fisheries as, for example in the North Sea case. Central to both types of delimitation have been the technical survey and legal rules, notably in establishing baselines and agreeing upon divisions of the continental shelf and its outer limits. These problems have been a major focus of the Law of the Sea negotiations since the 1950s.

From political and administrative points of view it may be that the most significant regional patterns are those which relate to contrasting structures and geographical scales of decision-making within several groups of national governments involved. First are the Western European states with a long history of maritime administration in which individual uses are generally linked administratively to cognate land departments. Interstate boundary problems are complex. This system prevails over much of the European seas, with isolated other examples such as Japan and New Zealand. Secondly are the large federal states, developed and developing, including the United States, Canada, Australia, South Africa, Brazil and India, which include the largest exclusive economic zones (EEZ), and in which state, rather than national boundaries may be of great significance. Thirdly are the centrally-planned states on the Soviet model, principally the USSR and eastern European states in the Arctic, Baltic, Black Sea and North West Pacific, together with Cuba. Fourth is China, (Figure 6.1).

The subdivisions within the smaller developing countries is less clear cut and it may be that basic cultural subdivision will be most appropriate, being respectively, the groups of states not already mentioned pertaining to Latin America; the Islamic world of North Africa and South West Asia; Africa south of the Sahara; South East Asia; the Pacific Island states, and the final, special case of Antarctica. In all these less developed countries (LDCS) maritime administrations are generally poorly developed, with substantial reliance in development, management and resource assessment placed upon

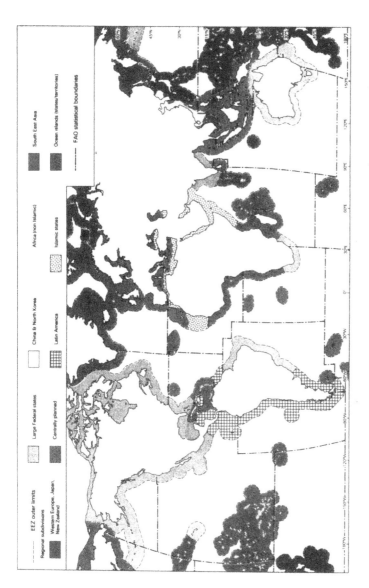

Figure 6.1. A regionalisation of the global Exclusive Economic Zone

bilateral public and private sector arrangements with industrial countries, and upon bilateral public and private sector arrangements with industrial countries, and upon the maritime-related agencies and bodies of the United Nations. By contrast, political disputes may be sharp to the point of violence in the case of boundaries between states.

Disputes apart, the political priority given to marine affairs is very variable. Although UNCLOS III has arguably greatly heightened awareness of marine affairs, the political profile of the sea is often low. This is the case even in industrial countries where the infrastructure of maritime administration is extensive.

THE REGIONALISATION OF MARITIME MANAGEMENT

The development of marine activities and management and related boundaries may be considered under six headings. The first three, shipping and navigation, strategic and fisheries uses have a relatively long history, and relate primarily to the use of the sea surface and water column, hence the long standing preoccupation with freedom of seas. The fourth area is the scientific study of the sea, where large-scale activity extends little over a century. Fifth is mineral exploitation, which belongs mainly to the last long wave of global economic development and has highlighted allocation problems and marine boundaries. Finally is environmental management wherein, perhaps, lies much of the future, as it will be central to the next long wave of global development. (1) Each management field may be briefly summarised in terms of objectives and boundary implications. (2)

The regulation and management of shipping and navigation has a comparatively long modern history, beginning perhaps with the British merchant shipping act of 1786. The primary objective has been safety, within which four major groups can be identified, namely, navigation aids, surveillance, emergency provisions, and regulation of shipping. The history of organisation has been characterised by the development of numerous bodies for each individual function, and the major international task since World War II has been that of global co-ordination of the rules and regulations by bodies such as the International Maritime Organisation, (IMO), International Association of Lighthouse Authorities, (IALA), and International Maritime Satellite

(INMARSAT). The legal framework throughout has been based on the freedom of the seas, an approach maintained in UNCLOS III. However, boundaries are administratively significant in three main contexts. The first is that of port operations; port traffic controls may extend well beyond legally defined limits of the port, or of the territorial sea for that matter, as in the case of oil terminals concerned with the saftey of loaded tankers. Secondly, surveillance and emergency provisions are related to coastguard and flight information regions which reflect overall shipping traffic patterns. Thirdly, in congested waters, are the designated traffic separation lanes, some of which lie beyond territorial waters, and are influenced by conflict with other sea users. These administrative boundaries often bear little or no relation to the political boundaries of states.

The concept of management in strategic affairs might seem to be rather alien, but it is a very important consideration as the primary objective of maritime strategic activity is the control of sea space. It is thus intertwined with surveillance, especially the coastguard group of activities, so that coastguards may become fully part of military organisation in wartime. The primary element in control is of course shipping and navigation and it is here that an important role for the political boundaries exists in defining the rights of naval forces on and beneath the sea surface, the freedom of the seas being again the basic legal framework. The strategic interest is historically also the principal driving force in the development of applied science, especially in monitoring the marine environment itself. That is why, for example, the Hydrographic Office and Meteorological Office in the United Kingdom are part of the Ministry of Defence. Naval testing areas are also a major, and sometimes exclusive use of sea space.

The fisheries, like navigation and strategy, are a water column activity with a long management history. However, it has had a much higher profile in management terms. Prior to UNCLOS III jurisdictional boundaries were a major means of fisheries regulation. To these have been added EEZs and EFZs in a number of cases, made possible administratively by the provision of catch statistics in certain areas. Jurisdiction and inter-state boundaries are thus used for the allocation objective, a purpose for which they are of limited real value. They are of even less value in realising the conservation objective, but have considerable potential use for the emerging fisheries planning objective concerned with integrating traditional fisheries management with regional planning objectives. The mobility of the resource is usually quoted as a major problem in the use of boundaries in fisheries management, but lack of adequate data may be a more important constraint. What is arguably required is a series of shifting boundaries related to individual stocks and patterns of fishing effort, and based on statistical monitoring regions which in turn are geographically based. The relation of fisheries management measures in the North Atlantic to the International Council for the Exploration of the Seas (ICES) grid system is an example of this, as is the relationship between the regional fisheries commission boundaries and the United Nations Food and Agriculture Organisation (FAO) statistical system based on the International Standard Statistical Classification of Aquatic Animals and Plants (ISSCAAP), (Figure 6.2). The current increase in economic pressures has elevated the priority of regional planning objectives, which in time will increase the significance of boundaries, but not those of traditional fishery limits.

Applied science also has a considerable history, extending from the early nineteenth century in the case of hydrographic surveying, and intimately related to the strategic and fisheries management objectives, despite the emphasis sometimes put on pure science, dating from the time of the Challenger expeditions. The most important single objective is the monitoring of environmental conditions, ranging from oceanographic characteristics to fish stocks; exploration, now especially of the seabed; special co-ordinated programmes usually concerned with collection of primary data and modelling; and environmental assessment of impacts, hazards and resources, where there is likely to be substantial social science inputs. The long-term implications of UNCLOS III are in the direction of restriction of scientific research by state interests in which political boundaries will become an obstacle. In the all important data dimension, however, the boundaries are primarily those based on latitude and longitude, and incorporating other technical and geographical considerations. Those of FAO and ICES have potential in various directions.

The field of mineral resource management is of course central to the present phase of boundary

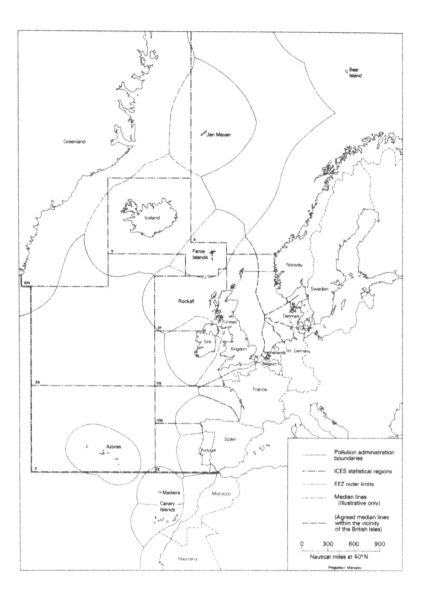

Figure 6.2. Maritime boundaries in the Northeast Atlantic

delimitation and indeed the whole 1982 Law of the Sea Convention, as far as the Area provisions go. Delimitation of the seabed provides a foundation for all the water column boundaries also, where and when considered necessary. Beyond the allocation objective, however, the detailed administration relies upon licensing systems mainly based upon latitude and longitude divisions which can be readily tied into scientific data bases. Mineral resources are of course state monopolies. As the controversial contesting of this gathers pace, state boundaries may become less significant, especially if private property rights were to be introduced on the sea bed.

Environmental management is the latest major sea use area. Most of this currently relates to pollution problems, but more emphasis is likely to be placed upon conservation and related research and recreation interests, as well as multiple use management problems. The pollution management boundaries adopted so far (as, for example, in the North East Atlantic) are related partly to navigation, and partly to fisheries and applied science water column boundaries which are most suitable (Figure 6.2).

MARITIME BOUNDARIES AND INTEGRATED MANAGEMENT

The pattern of boundaries in relation to ocean management reflects the separate organisational structures based on indiviudal use groups, while the complexity of management organisation and boundary development reflects the level of development. Thus the temperate oceans and seas of the North Atlantic and parts of the North pacific are most intensively used and managed. The remainder of the world ocean, including the ecologically sensitive polar and tropical seas, are less managed by comparison, although the management problems are considerable. The final theme of this chapter, therefore, is to assess the role of boundaries in relation to integrated management approaches which take account of the interrelationships among uses and organis-ations. This is considered in terms of over-all objectives, operations, organisation, data bases, and history of management.

The overall pattern of management objectives across all use groups falls into six major categories. (3) Four of these, safety, allocation, conservation and scientific research, are of long standing. The remaining two, regional development and multiple use management, have scarcely been explicitly thought of in the present context, except to a limited, but increasing extent in fisheries and offshore hydrocarbon exploitation. The construction of maritime boundaries has to date concentrated overwhelmingly on allocation among states; in the case of fisheries this has relied primarily upon jurisdictional boundaries, and for minerals upon allocation boundaries. For conservation (or, more widely, environmental control) and scientific research allocation and jurisdiction boundaries are generally unsuitable. Here the emphasis is upon the need for data collection boundaries which are primarily physical-geographical. Boundaries required for safety purposes are primarily related to patterns of economic activity, notably navigation and extraction of resources. This will also be the case of regional development and multiple use management objectives.

The operational aspects of sea use management consist of the practical functions involved. For example, in the case of fisheries management, there may be 10 to 15 separate tasks to perform in attainment of management objectives. These will have a variety of boundaries; thus surveillance may be related mainly to coastguard and strategic aerial patrol boundaries; allocation to both interstate and jurisdiction boundaries; and conservation to these and physical boundaries. When these operations are multiplied to cover the full range of use groups, a large number of different boundaries exist in an intensely used area where multiple use management approaches are required.

While the administrative organisation of management is based upon individual uses, the primary division is between governmental and non-governmental bodies; and the secondary distinctions are of scale among sub-national, national and inter- or supranational levels (Table 6.1). In both the perception of practical problems and in administrative organisation in both government and non-governmental spheres it may be that the sub-national local geographical scales and therefore boundaries are of the greatest importance. A good example is the detailed local limits and organisations in fisheries management.

The most important data bases are the scientific ones, and those concerned with navigation and resource extraction. The boundaries involved are those of research organisations and government

Table 6.1 Sea use management organisations classification

GOVERNMENTAL

International
    UN and UN Agencies and Bodies
    Regional treaty organisations
    Other regional organisations
    Other treaty organisations

Supra-national

National
    Government departments
    Agencies

Sub-national
    State/devolved governments and administrations
    Local authorities

NON-GOVERNMENTAL

Industrial organisations
    International
    National
    Sub-national

Special purpose organisations
    International
    National
    Sub-national

Information organisations
    International
    National
    Sub-national

departments involved in sea use administration. These boundaries are primarily based on latitude/ longitude, and secondarily on geographical consider- ations relating to the distribution of the phenomena of interest. In fisheries, for example, the primary subdivision globally is that employed by FAO which distinguishes polar, temperate and tropical oceans, and eastern and western oceans. In the North Atlantic, ICES sub-regions are based on physical divisions related to the distribution of fish stocks and fishing effort. Most continental shelf explor- ation and production licensing is based on latitude and longitude divisions and the distribution of mineral resources.

The handling of data draws attention to the fact that boundaries are not static. In construction of data bases there have been frequent changes in the nature and complexity of information including the geographical regions and time spans covered. Many physical boundaries, such as oceanographic and shore-line divisions, are dynamic, even over relatively short time scales. However, from a management point of view perhaps the most immediate interest centres round the long-term relationships between management on the one hand, and development on the other. There may, for example, be increased pressure for interestate boundary delimitation during long wave downswings in global development such as the present. During such periods emphasis tends to be placed on disaggregation and regionalisation of economic activity, rather than global integration. The true significance of the UNCLOS III Negotiations and the 1982 Law of the Sea Convention as economic exercises emphasise just such regionalisation, especially as regards the EEZ and, paradoxically enough, the Area provisions. The urge to create private property rights on the sea bed may be yet a further reflection of this. (4) Within large states, and the European Community, such regionalisation is an increasingly strong factor in, for example, the development of fisheries management regimes in the early 1970s.

CONCLUSIONS

The current period in history has been characterised by the strong emphasis placed upon the development of political boundaries among states, and jurisdictional boundaries reflecting the increased influence of states in marine affairs. It is necessary to distinguish, however, between industrial and developing countries respectively, and within these groups the patterns associated with western European states, large federal states, centrally planned states, the developing states of the continents, and small island states.

Marine management boundaries reflect the institutionalisation of management along the lines of individual uses, particularly in relation to navigation, strategy, fisheries, applied science, mineral exploitation and aspects of environmental management. These are only partly political, and many administrative boundaries relate to statistical and physical criteria. In particular, the pre-occupation with political and jurisdictional boundaries is really only suitable for allocation problems which involve fixed location activities, especially concerned with the sea bed and navigation. Water column activities, including fisheries, and scientific and environmental management necessarily rely upon other boundaries to be effective.

The increases in sea use activities and environmental impacts focus attention on the need for integrated management. Here the emphasis shifts from separately organised management structures to overall objectives, operations, integration or co-ordination of organisation, data bases and aspects of historical development. The boundaries required for each of these are use, physical and statistical boundaries related to global and regional development patterns. Such boundaries are different to those so emphasised in negotiations on the 1982 Law of the Sea Convention, negotiations strongly reflecting the current stage in economic and state development.

NOTES

1. For recent treatment of long waves see for example C. Freeman (ed) Long Waves in the World Economy (London, Frances Pinter, 1983).

2. H.D. Smith and C.S. Lalwani The North Sea: Sea Use Management and Planning (Cardiff, Centre for Marine Law and Policy, UWIST, 1984).

3. Ibid.

4. D.R. Denman, Markets Under the Sea, (London, Institute of Economic Affairs, 1984).

CHAPTER 7

COMMON FISHERY RESOURCES AND MARITIME BOUNDARIES: THE CASE OF THE CHANNEL ISLANDS (1)

Stephen R. Langford

Victor Hugo described the Channel Islands as '... pieces of France that have fallen into the sea and have been gathered up by England.'(2) Although lying close to the French coast, their allegiance to the British Crown can be traced back to the Norman Conquest.

The Islands comprise the Bailiwicks of the two largest islands of Jersey and Guernsey, each with their own dependencies (total land area of 195 square kilometres and a population over 130,000). The four largest islands (3) are self-governing, and as a whole, the Islands occupy an unusual constitutional position, being neither part of the United Kingdom, nor sovereign States, nor colonies. Instead, they are Crown possessions with their own legislatures, fiscal and legal systems, judiciaries and executives. The British Government is responsible for defence and international relations, whilst local affairs are looked after by the respective island governments. (4) Thus, the Islands' true constitutional status is one of semi-autonomy. (5) The result of this semi-independence is that Acts of Parliament do not apply to the Islands unless expressly named, or because they must apply by necessary implication. However, if the Islands wish, Acts may be extended to them on the approval of Her Majesty's Privy Council. Otherwise, the Islands exercise their own legislative powers. (6)

Legislation with respect to Channel Island fisheries has either been in the form of extended Acts of Parliament, or by agreements between the UK and France, binding the Islands only after their approval. It is argued here, that an anachronistic legal regime – as advantageous to France to preserve, as it is to the Channel Islands, (and more specifically Jersey), to remove - provides obstacles both to marine resource management and maritime boundary drawing in the area lying between the Channel Islands and the French coast.

THE CHANNEL ISLAND FISHING INDUSTRIES

Fishing in the Channel Islands has existed since earliest times: indeed, between 1100 and 1300, it was Guernsey's main source of wealth. (7) However, increased competition, first from Newfoundland, and then from England, reduced fishing to the position of a highly localised industry from the beginning of the 16th Century. In the late 19th Century, Hornell refers to 'competition at a dis-advantage with larger and better equipped French craft', (8) which not only poached, but could stay at sea longer; whilst Holdsworth describes a largely inshore fishery, more important in Guernsey than Jersey. (9) Between the World Wars, pollution, poaching, lack of labour and cheap imports led to decline, (10) so that, in 1953, Hooke wrote that Guernsey's fishing industry was dead. (11)

However, during the 1960s a market for shellfish developed rapidly in Europe and the UK, which benefited islands favourably endowed with shellfish stocks, especially lobsters and crabs. Since then, the fishing industries have returned to prominence. During the 1970s, whilst the traditional Channel Island industries of horticulture and tourism underwent a difficult transitionary period, the fishing industries of Guernsey and Jersey grew considerably. For example, between 1970 and 1981, Guernsey's fish catch increased from 250 tonnes worth £91,000, to 2,900 tonnes worth £3 million; whilst the number of Guernsey fishermen increased from 109 in 1960 to 900 in 1981. The number of fishing boats registered, increased from 60 to 566 over the same period.(12) Whether such expansion can be maintained depends, not only upon the management policies adopted by the Islands, but also, more particularly for Jersey, upon the actions of the French. This is because of the existence of the so-called mer commune.

THE ORIGINS OF THE MER COMMUNE

An Anglo-French Fisheries Convention signed in Paris on August 2, 1839, (13) defined and regulated the limits of oyster and other fisheries off British and French coasts. It arose out of numerous

complaints by British and French fishermen over each
other's actions, in the absence of any legislation
establishing the outer limits of either of the two
States' respective territorial seas, and hence,
their exclusive fisheries' limits. The most bitter
French complaints were directed against the British
fishermen dredging for oysters off the French coast.
The British fishermen, in turn, complained the
French were interfering with oyster dredging up to
fifteen miles off the French coast. In 1837 an
Anglo-French commission was appointed to look at
these complaints, and particularly, to discover and
define the limits within which each State's
fishermen 'should be at liberty to fish for oysters
between Jersey and the neighbouring coast of
France.' (14)

The oyster beds to the north-west of the
Chausey Islands had been fished extensively from
1797 by French, English and Channel Island
fishermen. (15) The French objected to British
trawlers dredging within two marine leagues of their
coast, and were determined to claim the oyster beds
as the exclusive preserve of French fishermen, (16)
despite the fact that they lay outside the three
mile limit claimed by Britain to be the law of
nations. (17) By Article I of the 1839 convention a
'special régime' area exclusively reserving the
oyster fishery for French fishermen was established.
This stretched for seventeen miles from Cape
Carteret in the north to Port Meinga in the south,
well in excess in some places of the three-mile
limit, and at one point fourteen miles offshore. (18) Its outer limits were defined by
straight lines forming the French A-K line (Figure
7.1), the lines being drawn both to follow
particular oyster banks, and to provide suitable
landmarks for fishermen to observe. (19) Indeed, at
that time, it was usual to recognise possible
'exclusive' rights beyond the limits of the
territorial sea when referring to oyster fisheries,
due to the fisheries' stationary nature making it an
exhaustible resource, and on the basis of long
use. (20)

The Convention went on to define a non-existent
oyster fishery within three miles of the low-water
mark of Jersey, reserved exclusively for British
fishermen (Article II). Between the limits set out
in Articles I and II the oyster fishery would be mer
commune to both Britain and France (Article III).
Within three miles of the 'whole extent of the
coasts of the British Islands', British subjects

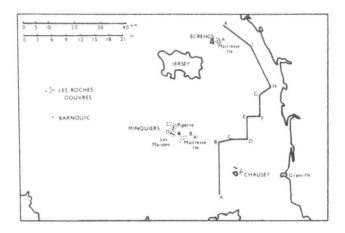

Figure 7.1. The areas delimited by the Articles of the 1839 UK-France Fisheries Convention (Representational)

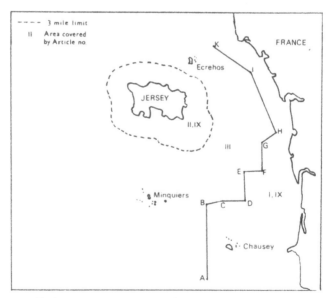

Figure 7.2. The concession areas of exclusive fishing rights within the Minquiers and the Ecrehos under the 1951 UK-France Fisheries Agreement (Representational)

would have exclusive rights of fishery (Article IX). Similarly, French subjects would have exclusive rights of fishery within three miles of the French coast, with the exception of the 'special régime' area laid down in Article I, where the limits of the oyster fishery were to apply.

The 1839 Convention was thus a confusing document dealing both with limits to the local oyster fishery, and to the general fishery on all coasts. Under it, the French oyster fishery was defined differently from the French area of exclusive general fishery; and the Jersey and French 'exclusive' oyster fishery zones of Articles I and II were defined, in the Jersey case, by describing the baselines from which the zone was to be measured, ie the low-water mark, and in the French case, by straight lines delimiting its outer boundary. The waters between the outer limits of each exclusive zone were a common oyster fishery under Article III. However, by Article IX, for general fisheries purposes, although the exclusive zones were defined as coterminous with the three-mile territorial sea measured from the low-water mark, the French were granted exclusive rights of general fishery within the special régime area, set out as the exclusive French oyster fishery under Article I.

Fulton relates how virtually all of the 1839 Convention's 'elaborate regulations' turned out to be either unworkable or came to be disregarded. There was also much disagreement as to what actually were 'the seas lying between the British Islands and France', to which these regulations applied. (21) Thus, on November 11, 1867, a replacement convention was signed, which revised the 1839 Convention supposedly in the common interest of both States' fishermen. (22) Under Article I, the exclusive fishing limits of both States were set out as under Article IX of the 1839 Convention, with the same exception for the French 'special régime' area defined by the A-K line, delimited as in 1839 (Article II, 1867). As in 1839, this was also to be exclusive French fishery, although outside the 3 mile limit in places. Under Articles III and X of the 1867 Convention, fishing of all types was permitted outside the respective exclusive fishing limits, with the exception of a closed season for oysters, established for the English Channel.

Therefore, the 1867 Convention made no mention of the common oyster fishery established under Article III of the 1839 Convention. Indeed, under

Article XLII (1867) this mer commune area would cease to exist once the new convention entered into force. Unfortunately, although the 1867 Convention was ratified on January 14, 1868, it never entered into force, because of French objections to some of its provisions. (23) As a result, the 1839 Convention remains in force today, and has created problems, both for boundary drawing and resource management, in the waters between the Channel Islands and France.

FISHING LIMIT PROBLEMS CONCERNING THE MINQUIERS AND ECREHOS

The Ecrehos and the Minquiers are two groups of isolated rocks and islets, only two or three of which are habitable. The Ecrehos lie 3.9 miles north-east of Jersey, and 6.6 miles from the French coast, while the Minquiers lie 9.8 miles south of Jersey, and 8 miles from the French Chausey islands. (24) The French had not questioned their status as dependencies of Jersey, until her fishermen began to take an interest in their waters, leading in the 1880s to French sovereignty claims in respect of the islets. The French view was that, under Article III of the 1839 Convention, the waters surrounding the Ecrehos and Minquiers were mer commune. Neither group had been mentioned in 1839, as the French had not questioned their British sovereignty, but now the French claimed they were 'neutral' because they were not mentioned by the Convention.

In 1857 the Jersey States had declared the Ecrehos part of the parish of St Martin, and in 1883 the British reaffirmed their status as dependencies of Jersey, reminding the French that they had not objected to the Ecrehos' inclusion as part of the port of Jersey in 1875; nor had they attempted to alter the 1839 fishing limits to include the two groups when the 1867 Anglo-French Fishery Convention had been agreed. Jersey fishermen petitioned the States to protect their exclusive rights of general fishery within the three-mile territorial sea, to which they were entitled under Article IX of the 1839 Convention. The French countered, that if the limit of the British exclusive fishery were taken at three miles from the Ecrehos, instead of three miles from Jersey, it would entirely absorb the mer commune provided for by Article III (1839). This French contention failed to recognise, (or chose to ignore), that their argument applied only to the oyster fishery, whose limits were fixed by Articles I-III, and not to the general fishery limits set out under Article IX as extending 'along the whole extent of the coasts of the British Islands'. The only way in which the position could be different, therefore, was if the French were to be found to be the sovereign power of the Minquiers and the Ecrehos. Thus, France claimed sovereignty over the Ecrehos in 1886, and over the Minquiers in 1888.

The French sovereignty claims were dubious, in view of the fact that France had never exercised any sovereign duties over the Minquiers and Ecrehos, nor had she sought to do so before her fishermen took an interest in their waters. Indeed, France alternated between claiming the islets as her own, and maintaining their neutrality, based on her view that they lay within the mer commune area. Johnson notes:

> The French contention involved not only treating land as sea but also drawing altogether excessive conclusions - relating to such an important matter as sovereignty - from a mere oyster fishery provision. (25)

However, before World War II, UK policy had been to allow French fishermen to fish within the islets' claimed three-mile territorial sea, whilst, at the same time, asserting British sovereignty. Thus, equality of fishing opportunity, and hence a mer commune arrangement, could be said to exist, irrespective of the territorial sea claims.

Neither areas were fished seriously by the French until after World War I, when Jersey fishermen grudgingly tolerated French daytrippers' picnic parties and the islets' use as a landing stage by French fishermen. However, during the latter part of World War II, the German Occupation of the Channel Islands meant Jersey fishermen were unable to fish these waters, whilst the French, having been liberated, were able to. (26) After the war, Jersey fishermen encountered hostile French fishermen, who claimed that the waters, especially around the Minquiers, belonged to France. (27) As a result, the International Court of Justice (ICJ) was asked to determine whether Britain or France held sovereignty over the Ecrehos and Minquiers, 'insofar as they were capable of appropriation'. (28) However, the French only agreed to arbitration on condition that the fishery question was settled first.

Article I of the UK-France Agreement of January

30, 1951, reaffirmed the 'common rights' concept, interpreting the 1839 Convention as conferring upon British and French nationals 'equal rights of fishery' between the limits of three miles on the coast of Jersey and the limits of French fisheries in Granville Bay, ie the 'special régime' area, (the A-K line having been slightly modified by the UK-France Agreement of December 20, 1928). (29) The only exceptions were four small circular areas within the Minquiers and the Ecrehos, for which exclusive fishery rights concessions could be granted, according to the award of sovereignty. These zones were: A - a radius of one third of a mile from a beacon on Maitresse Ile in the Ecrehos (Article II); B - a radius of half a mile from a flagstaff on Maitresse Ile in the Minquiers (Article III); C - a radius of half a mile from a beacon on a specified rock amongst the Pipette Rocks in the Minquiers (Article IV(i)); and D - a radius of half a mile from a beacon on a specified rock amongst the Maisons rocks in the Minquiers (Article IV(ii)). Within each zone certain named rocks were to be included or excluded. Zones A-C could be claimed by Britain if she were found to have sovereignty; zone D could be claimed by France if she were sovereign (Figure 7.2).

In effect, the Agreement made common, apart from these restricted concessions, any fishing rights which would otherwise have been territorial and exclusive as a result of the sovereignty arbitration. The French were thus permitted to pre-empt any of the benefits which would accrue if, as happened, Britain was awarded sovereignty. It virtually said that, for fishing, if the French case was rejected by the ICJ, the matter would nevertheless be as if it were accepted.

Furthermore, an important mistake was made in the omission of the qualification 'oyster' to the fishery disputed in the 1951 Agreement. Thus, the mer commune area under Article III of the 1839 Convention, which referred to a common oyster fishery, became a common area of general fishery, and so included the territorial waters of the Ecrehos and the Minquiers. Moreover, by referring not to the oyster area of 1839/1928 but to the all-coast exclusive fishing areas of Article IX, the Agreement raised the question of a western boundary to the mer commune without defining it; for prima facie the 1951 regime applied to any area lying between the Jersey and French exclusive fishery zones on all coasts, not just between the Jersey exclusive fisheries zone and the French counterpart, defined by the A-K 'special régime' line in the eastern part of the Bay of Granville. In 1953, the ICJ unanimously awarded sovereignty of the Minquiers and Ecrehos to Britain. (30) Britain produced numerous historical documents to support its sovereignty claim, which, though not mentioning the two groups by name, could be taken to refer to them by implication. This they backed up with evidence of acts manifesting continuous sovereignty over the islets, eg the erection of customs' houses, property sales, inquests and taxation, and evidence that since the early 19th Century, both reefs had been inhabited by Jersey residents. The French in turn, repudiated any historical evidence not expressly naming the two groups, claiming that they had become French possessions upon the 'dismemberment' of the Duchy of Normandy in 1204. They also pointed out, that since 1861, France had assumed sole responsibility for providing and maintaining the lights and buoys on the Minquiers without British objection. Although the ICJ's French judge, M. Basdevant, held that Jersey's unilateral expression of her convictions of her own sovereignty was not sufficient proof of actual sovereignty, (31) even he accepted that Jersey had exercised 'greater and more continuous activity' over the islets than France. (32) The judgement thus confirmed British claims to sovereignty of the Minquiers and the Ecrehos.

Taken together, the 1951 Agreement and the 1953 sovereignty award, gave the UK the right to grant exclusive fishery rights to Jersey fishermen in zones A-C. This she wished to do to assert British sovereignty, in recognition of the importance attached to effective possession by the ICJ. However, the UK also wished sovereignty to be exercised leniently and loosely, and suggested that unless there was some real advantage to be gained by granting exclusive rights to Jersey fishermen, the latter might refrain from any such claims. Only one fisherman applied, thereby confirming the value of the concessionary zones to be small. Thus, the application was refused and the rights to grant exclusive rights remain reserved.

More importantly, the French subsequently argued that by signing the 1951 Agreement and accepting concessionary zones, the UK had renounced in advance of the sovereignty award, the existence of territorial seas around the Ecrehos and Minquiers, and therefore their waters were to be considered part of the 'high seas'. In reply, the UK held that the 1951 Agreement concerned fishing, and did not mention the territorial sea, which though usually co-existent with fishery limits, also involves rights to minerals and other non-living resources, plus jurisdiction in general. As the territorial sea includes obligations, as well as rights, it is an inalienable appurtenance. Therefore, islands found by the ICJ in 1953 to have been part of Jersey long before 1951, could not be deprived of a territorial sea by the absence, in a bilateral fisheries agreement, of a provision specifically mentioning territorial seas. Furthermore, the French decision to cease maintaining the buoys at the Minquiers after 1956 was suggestive of a recognition of territorial sea sovereignty. (33)

Jersey's rights and obligations within territorial seas, defined as a three-mile circle centred on Pointue a Sablons in the Minquiers, and a two-mile circle centred on Maitresse Ile in the Ecrehos were stated in the Jersey Sea Fisheries Law of 1962 (Figure 7.3). The Ecrehos two-mile territorial sea was a concession to the French, recognising their special position in relation to the two reefs, and also the ICJ judgement which left the extent of the sovereignty awarded an open question, bearing in mind the difficulty of determining the area of land capable of appropriation. Thus, for the purpose of fishery regulation, it was decided that the existence of the reefs themselves was more important than the areal extent of sea the reefs enclosed; hence, the more compact Ecrehos should have a smaller two-, instead of three-mile territorial sea, a decision which is probably unique in the world, and suggestive of confusion of territorial rights with fishery zones. (34) Indeed, the decisions relating to the Minquiers and Ecrehos suggest sovereignty may not indicate territoriality, and that the territorial sea may not be the sovereign State's exclusive fishery.

DEVELOPMENTS CONCERNING THE MER COMMUNE SINCE 1964

Further problems arose out of the London European Fisheries Convention of March 9, 1964, which led to the Fishery Limits Act (1964), extending British fishing limits to an exclusive six mile zone, plus a further six mile 'outer zone', in which habitual fishing rights would be observed. This Act was extended to the Channel Islands, but the French refused beforehand to relinquish their

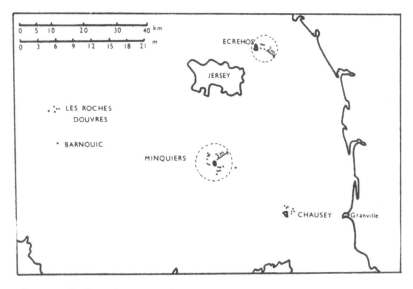

Figure 7.3. The territorial seas of the Minquiers and the Ecrehos as described in the 1962 Jersey Sea Fisheries Law (Representational)

fishing rights in the mer commune, as laid down in the 1951 Fisheries Agreement. Therefore, on April 10, 1964, in an Exchange of Notes between the UK and France, (35) it was agreed that, notwithstanding this Act, it 'would not affect the special regimes established between the two countries concerning Granville Bay and the areas of the islands of the Minquiers and Ecrehos group.' The provisions contained in the 1839, 1928, and 1951 agreements would continue to be applicable to French and British fishermen, and would be transferred to a bilateral agreement, to be concluded as soon as possible.

In 1965, Jersey fishermen's fears for their livelihood under the mer commune arrangement, prompted the UK to open negotiations with France. During discussions, the French asked for clarification of the western boundary of the mer commune, suggesting the L'Etac de Sark line. Verbal agreement by Jersey's representatives resulted in not only Jersey's eastern coast (enclosed under the 1839/1928 Conventions), but all its coasts being enclosed by the agreed concept of a mer commune, as

99

the mistake made in 1951 had made possible. Therefore, the L'Etac de Sark line was written into the UK Note of February 18, 1965 as the north-western limit of the Bay of Granville's 'special régime'; ie of the mer commune, even though the 1964 Exchange of Notes had referred only to the narrow waters between Jersey and the French coast, and not to any western boundary.

The draft bilateral agreement proposed under the 1964 Exchange of Notes, was drawn up in 1968. Its terms were basically an adaptation of the mer commune concept, including provisions for equity and conservation. Article 2 provided for an area of common fishery rights for French and British fishermen; Article 4 delimited an area known as Beasley's Bulge, where fisheries were common except for lobsters, which were to be exclusively reserved for British fishermen. (36) Article 5 set aside an area lying within Jersey's 3 to 6 mile belt, that could be used for twenty years by French fishermen long lining, trawling or dredging; and Article 6 gave the French permission to long line for a period of five years, in an area within Jersey's three-mile territorial sea, north of the Paternosters (Figure 7.4). However, this draft agreement was never ratified, presumably because its terms were regarded as unfavourable to French interests, the French standing to gain most from the retention of the mer commune because of their more rapacious fishing methods.

Since 1968, Jersey has claimed habitual fishing rights in three areas of the French outer belt: north and east of the Roches Douvres, north of Cherbourg, and three miles north of the Ecrehos (Figure 7.5). The French have tolerated Jersey boats in the Cherbourg 6 to 12 mile belt and around the Roches Douvres (where French fishermen are largely inactive), but not UK or Guernsey boats. Guernsey has also claimed habitual rights off the Roches Douvres and east of the Banc de la Schole, but these have been ignored because of the French refusal to settle the Guernsey claims before concluding a settlement over the mer commune. This expresses a French preference to look at the matter of the Channel Island waters as a whole. This is especially annoying to Guernsey, because French boats are able to poach in their outer belt up to the L'Etac de Sark line, claimed by France as the mer commune boundary. This line is in fact two miles closer to Guernsey than the median line between the two Bailiwicks, and thus, part of Guernsey's outer belt.

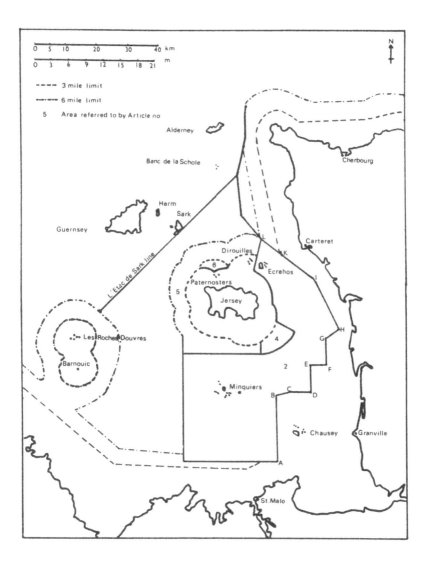

Figure 7.4. The 1968 draft UK-France Fisheries Agreement (Representational)

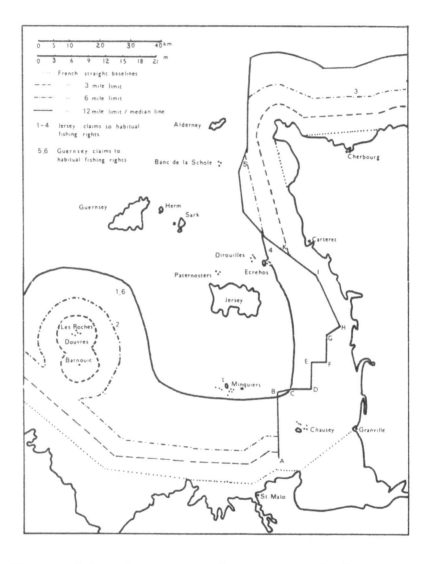

Figure 7.5. The areas of Jersey and Guernsey habitual fishing right claims in France's 6 to 12 mile outer belt (Representational)

Common Fishery Resources: the Channel Islands

A Frenchman found fishing in this disputed area in 1978 believed he was entitled to fish there. The small fine imposed on him, (£200 compared to a maximum of £50,000), was accompanied by the admission that French fishermen were ignorant of the limits through no fault of their own. (37) Two years later, another Frenchman had his case of alleged poaching dismissed, because the clarification of fishing limits to the south of Guernsey, requested by France, had still not been given. (38) The continuation of the mer commune thus impinges upon Guernsey, as well as Jersey waters.

During the 1970s, the 'beaches doctrine', which would have allowed EEC member countries free access to each other's territorial waters, threatened the Channel Island fishing industries. This doctrine would have come into effect on January 1, 1983, had not a Common Fisheries Policy (CFP) been agreed and signed on January 25, 1983. The CFP provided for existing six-mile exclusive belts, and habitual rights within the 6 to 12 mile belt for member States.

The ambiguous status of the Channel Islands as Associate members of the EEC leaves them outside of the CFP, which constitutionally they could not adopt, without, as a consequence, having to adopt other EEC regulations necessitating changes in the Islands' constitutional status. However, the existence of the mer commune virtually demands that the Islands enact their own legislation parallel to the CFP, to keep in line with the UK and France in matters such as regulations designed to conserve fish stocks. More significantly, two Jersey vessels were arrested for alleged illegal fishing in the Irish Sea in 1982. The Irish claimed that as the Channel Islands were not in the EEC, they were not eligible to fish EEC waters. If this was applied by the UK and France, the consequences would be disastrous, two-thirds of Jersey's fish coming from EEC waters. Reliance upon the mer commune would, therefore, increase to a level which fish stocks could not maintain, except in the unlikely event of the French agreeing to allow Jersey sole use of the present mer commune area.

A REPLACEMENT BOUNDARY FOR THE MER COMMUNE?

The 1839 fisheries' régime is undoubtedly a legal anachronism, which was never intended to be perpetual. However, it was reserved, protected and given new life by the 1964 Exchange of Notes. Until recently, the removal of the mer commune would have threatened more than just the uncovering of the full effect, in respect of exclusive fishing rights, of France's 1967 declaration of a twelve-mile territorial sea on straight baselines. (39) Ninety-five per cent of Jersey's fish was exported to France, and Jersey could not afford French reprisals (eg import bans), in the event of the mer commune being dismantled. This was far more significant than any expert legal opinion, which held that the mer commune would not stand up in a court of law if challenged. (40) Jersey fishermen preferred to strengthen their claims for habitual rights within French exclusive fishing areas, in anticipation of the mer commune being fished out, rather than challenge its existence in a court of law. However, fears of French reprisals, in the event of Jersey successfully revoking the mer commune, have now been removed, as Jersey now exports the majority of her fish direct to Spain and Portugal. (41)

Under the mer commune arrangement, Jersey's exclusive fishing limits are frozen by the 1951 Agreement to three miles (ie she cannot extend her territorial sea). Off her south and north-east coasts, Jersey is bound to allow French fishermen free access to the Ecrehos and Minquiers, the former of which lies just outside the three-mile limit. It is clearly to the French advantage to retain the mer commune, as their own coastal fisheries are exhausted. Moreover, French fishermen are legally able to fish Jersey waters up to the three-mile limit, whereas Jersey fishermen fish within French waters on suffrance, a privilege that can easily be removed. Indeed, within the mer commune area, the French legally use more destructive fishing methods than their Jersey counterparts, and operate less stringent regulations on undersized fish and net mesh sizes. It appears, therefore, that one possible alteration to the current situation would be a division of the mer commune by an east-west line running through the Minquiers, as the French tend to concentrate their activities to the south, and Jersey fishermen to the north, of this line. To do this, however, would again call into question the sovereignty of the Minquiers.

Professor Bowett, legal adviser to the Jersey States, (42) apparently recommends that Jersey extend her territorial sea to her legal entitlement of twelve miles drawn on straight baselines, (to mirror the 1967 French declaration), in anticipation

of a territorial sea boundary line between the
Jersey and French coasts. In Jersey's case, the
Minquiers and Ecrehos would be linked to the main
island for the purposes of drawing straight
baselines, which would substantially increase
Jersey's area of territorial sea. It is not clear
whether the linkage by straight baselines envisaged,
is to be based on the 1982 Law of the Sea
Convention's rules for straight baselines (Article
7), or its rules for archipelagic baselines (Article
47). Neither group of islets appears to fulfil the
condition of being 'a fringe of islands along the
coast in its immediate vicinity' (Article 7(1)), nor
would the baselines drawn, follow 'the general
direction of the coast' in such a way, that the sea
areas lying within the lines would be 'sufficiently
closely linked to the land domain to be subject to
the regime of internal waters' (Article 7(3)).
However, Jersey, the Minquiers, and the Ecrehos,
might be said to be an archipelago whose 'islands,
waters and other natural features form an intrinsic
geographical, economic and political entity, or
which have been historically regarded as such'
(Article 46(b)); in which case, it is difficult to
see how archipelagic baselines thus drawn, would not
conform to Article 47's rules.

In principle, a territorial sea boundary would
define the exclusive fishery limit, and thus remove
the mer commune, unless some special fisheries
arrangement were agreed. However, it will be
necessary to clarify the fishing and baseline
position before the Channel Islands/France boundary
line is delimited. What respective French and
Channel Island jurisdictions this boundary line will
delimit is by no means clear. The most logical
solution would be a territorial sea boundary
separating overlapping 12 mile claims, but at
present the Channel Islands territorial seas are
only 3 miles, although there is a 12 mile fishery
limit around the Channel Islands, established in
conformity with the European Fisheries Convention of
1964. However, this fishing limit is inapplicable in
the mer commune area, as stated in the 1964 Exchange
of Notes. The UK adheres rigidly to a 3 mile
territorial sea, and, notwithstanding the Channel
Islands non-inclusion in the UK, it is unlikely that
they will be allowed to unilaterally extend their
territorial seas to 12 miles without UK
approval. (43)

In 1977, the Court of Arbitration in the UK-
France Continental Shelf Case (44) noted that the UK

and France had provided arguments concerning the delimitation of a continental shelf boundary between the Channel Islands and France. (45) It further noted that this 'continental shelf' boundary '... must traverse over almost its whole length waters either claimed by France as part of its territorial sea or by the United Kingdom as part of its actual or potential territorial sea and of its existing fishing zone.' (46) The phrase 'over almost its whole length', could be taken to indicate knowledge of the mer commune, and indeed, the French Agent held that it was possible that the competence of the Court to delimit the boundary could '... be denied by one or the other Party, in at least an important portion of the area' between the Channel Island and French coasts. (47) This 'important portion of the area' may refer to the mer commune area, but at no point in its judgement does the Court give any explicit indication that it knows of the mer commune's existence. Rather, the Court was concerned about its competence to delimit, what it perceived to be, a territorial sea boundary in this area. Knowledge of the mer commune could not have allowed the Court to perceive that it was being asked to delimit a territorial sea boundary, given that, in effect, the mer commune is 'high seas'.

The Court, therefore, asked the parties what competence had been conferred upon it to delimit an apparent territorial sea boundary, given that Article 2(1) of the Arbitration Agreement had asked it to delimit a continental shelf boundary. (48) The French Agent's response was that the Court could delimit the continental shelf, only if the boundary line was located beyond the limits of the territorial sea of one of the Parties or coincided with those limits. (49) This implied that a continental shelf boundary delimitation was possible, given that the Channel Island territorial sea was set at 3 miles. The UK's Agent's reply was more explicit: noting that at present the UK claimed only a 3 mile territorial sea, he held that the boundary claimed by the UK in this area was, with the exception of one small segment off the Ecrehos, a continental shelf boundary, notwithstanding the fact that parts of it might be coincident with the French territorial sea boundary. (50)

The Court eventually decided that it did not have the competence to delimit the boundary between the Channel Islands and the French coast, because France had indicated its competence might be disputed in one important area, which it regarded as territorial sea, and the UK as continental shelf. This was a convenient finding. The Court was thus able to side-step the disputes concerning relevant basepoints, allowing it to conclude that:

> In narrow waters such as these, strewn with islets and rocks, coastal States have a certain liberty in their choice of basepoints; and the selection of basepoints for arriving at a median line in such waters which is at once practical and equitable appears to be a matter peculiarly suitable for determination by direct negotiation between the Parties. (51)

The pleadings in the UK-France Continental Shelf Case have never been published, but it is understood that the French Memorial to the Court repeated the French assertion that, under the 1951 Fisheries Agreement, the UK had relinquished its rights to territorial seas around the Minquiers and Ecrehos, and thus the islands were part of the high seas. To this they added, that the principle of natural prolongation did not justify their being taken into account in negotiations over the Bay of Granville median line, (52) as they formed part of the natural prolongation of the French continental land mass. However, the French stressed that the Roches Douvres and Chausey Islands could be taken into account, as France had always asserted her sovereignty over their territorial seas, and fully exercised her powers regarding fishing in their coastal waters.

In turn, the UK Memorial asserted UK rights to territorial seas around the Minquiers and Ecrehos, (53) and pointed out that at one stage in negotiations, the French had offered to ignore the Chausey Islands in the delimitation, if the UK ignored the Minquiers. The fact is that, without this concession, a part of the Minquiers, despite the 1953 ICJ ruling, would become French continental shelf. Moreover, there would be some inconsistency in taking account of the Roches Douvres, but not of the Minquiers and Ecrehos. The French proposition to take account of the bare rocks of the Roches Douvres, but not of the populated Chausey Islands, probably has something to do with the fact that the latter lie within the A-K 'special régime' area, and their non-inclusion is thus no concession whatsoever. However, the Court's judgement provides no allusion to these <u>mer commune</u> related issues. It merely reports that if the Minquiers are not accepted as a basepoint, the UK disputes the use of the Roches Douvres as basepoints; and if the Minquiers' use is acceptable, France asserts a right to use the Chausey Islands as basepoints. (54) Thus, the judgement's focus on the basepoint disputes appears completely ignorant of, or else is purposely divorced from, the mer commune's continued existence.

The judgement does reveal, however, that the UK and France are agreed that, 'in principle', a median line will divide these narrow waters. (55) Jersey is thus charged with the responsibility of changing the mer commune arrangement to allow an extension of her territorial sea to twelve miles, which, as a consequence, would require a boundary line delimitation. Until this happens, the de facto median lines recognised by France and the Channel Islands will continue to differ, because of the disputes over relevant basepoints (Figure 7.6).

The 1977 UK-France Continental Shelf Case established 12 mile enclaves to the north and west of the Channel Islands, and it was rumoured in 1982, that enclaves to the south and east were being mooted as the solution to the problems of the Bay of Granville, though this was denied. (56) However, Jersey hopes to reopen negotiations concerning its maritime boundary with France, partly because here dependence on French fish markets has been removed; and partly because of increased activity of French vessels on the 'Jersey side' of the mer commune, compared with that on the 'French side', which has been thoroughly overfished. A recent air survey by UK authorities recorded 147 French, and 36 Jersey and Guernsey vessels, fishing the mer commune. Conservation is thus a good argument for a median line solution. (57)

CONCLUSIONS

Discussing the 1951 Fisheries Agreement and the 1953 sovereignty award, Alexander states that:

From the standpoint of generic concepts it would seem that this case is to a large extent unique, with few 'principles' of offshore claims which might be related to other areas ... But geographers, seeking to describe the variable character of the earth's surface, have no cause to ignore the unique, since it

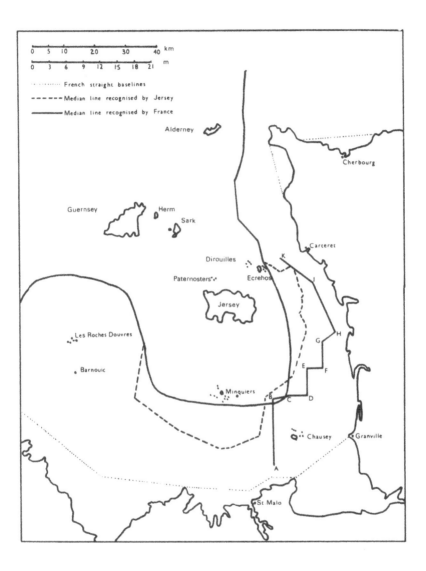

Figure 7.6. The _de facto_ median lines recognised by France and the Channel Islands (Representational)

too gives character and meaning to the specific area in which it occurs. (58)

Though to a large extent the problems of the mer commune may be unique, the problem of interpreting historical boundary agreements, in order to establish unambiguous maritime jurisdictional limits, is not uncommon. Unfortunately, attempts to clear up ambiguities in the 1839 UK-France Fisheries Convention have created their own problems.

Significant mistakes were made in the removal of the word 'oyster' from the 1951 Fisheries Agreement; in allowing a fisheries agreement to be concluded prior to the sovereignty decision concerning the Minquiers and Ecrehos; and in recognising the L'Etac de Sark line as the boundary of the mer commune. Thus, France has been allowed to gather important legal holds binding the Channel Islands to an anachronistic regime that France has no reason to remove. Although negotiations concerning the mer commune have occurred intermittently over a long period of time, the Channel Islands always negotiate from a position of weakness, both legally and politically. Maintenance of the status quo is clearly to France's advantage, and it therefore seems unlikely that the mer commune will be negotiated away. It seems equally unlikely, in the absence of a negotiated agreement, that the UK and France would be willing to incur the costs of an arbitrated settlement, concerning a relatively small area of little significance to either of them, notwithstanding the area's importance to Jersey. (59) It is also doubtful whether Jersey would be able, or wish, to incur the costs of litigation herself. (60)

Therefore, the removal of the mer commune seems possible only in the highly unlikely event of hydrocarbon discovery in these narrow waters, (61) or when the area is thoroughly fished out. The oyster fishery in the Bay of Granville, which the 1839 Convention had sought to regulate, was exhausted by 1921, destroyed by a combination of disease, overfishing, and damage caused to seabed spawning grounds by dredging. (62) Today, ever increasing fishing effort in the mer commune is beginning to be reflected in lower catches per unit effort, usually a sign of overfishing. Thus, there is the future threat of the Channel Islands' fishing industry reverting to a pre-1960's cottage industry status. As sedentary shellfish stocks remain of paramount importance to continued expansion of the Channel Island fishing industries: 'The maximum control over the largest possible area of fishing resources is an obvious pre-requisite for the future development of the fishery.' (63) The existence of the mer commune clearly frustrates such development. Whittlesey astutely observed that:

> Legal systems are images of the regions in which they function, sometimes faithful and sometimes distorted ... Because humanity occupies its habitat dynamically, laws tend to be outmoded. When this occurs they are occasionally revoked, sometimes they are disregarded, usually they are given new meaning. Always there is a lag between the reason for change and its legal accomplishment. (64)

In the case of the offshore geography of the Channel Islands, this time lag dates from the exhaustion of the oyster fishery in 1921. The agreements of 1928, 1951, and 1964, have perpetuated an 1839 Convention designed for one specific purpose: the regulation of the oyster fishery of the Bay of Granville. The legal regime is outmoded, it should be revoked, but it is hard to disregard. It has been given new meaning, and there is a reason for change, but its legal accomplishment is still awaited.

NOTES

1. This paper is a revised version of part of The Offshore Geography of the Channel Islands. (B.A. dissertation, University of Oxford 1984). I particularly acknowledge the assistance of Capt. F.W. Jeune (Assistant Harbourmaster of Jersey, retd) in its preparation.

2. Victor Hugo Les Travailleurs de la Mer (1886), p.XXI.

3. Jersey, Guernsey, Alderney and Sark.

4. These are the respective States of Deliberation of Jersey, Guernsey, and Alderney, and Sark's Chief Pleas.

5. Sir John Loveridge, The Constitution and Law of Guernsey (1975), p.1, (Guernsey, 1975).

6. Ibid, pp.3, 19.

7. See T.W.M. de Guerin 'Feudalism in Guernsey' Transactions of La Société Guernesaise 1909, (1909), 1-25, cited in M.R. Dunn, C.G. Askew and D. Whitmarsh The Development of the Guernsey Fishing Industry, p.1, (Marine Resources Research Unit, Portsmouth Polytechnic: Unpublished, 1977).

8. See J. Hornell 'The possibilities of fishery importance in Jersey' Journal of Marine Zoology and Microscopy, II (1897), 73-94, cited in Dunn et al op.cit., p.8.

9. See E.W.H. Holdsworth Deep Sea Fishing and Fishing Boats, (1874), p.216, cited in Dunn et al op.cit., p.9.

10. See Victor Coysh 'Guernsey fishing boats' Transactions of La Societe Guernesaise, XIV (1947), p.133, cited in Dunn et al op.cit., p.10.

11. See W. Hooke The Channel Islands (1953), p.119, cited in Dunn et al op.cit., p.13.

12. Dunn et al op.cit., p.18. The number of full-time fishermen was 43 in 1960, and 250 in 1981. See also States of Guernsey Sea Fisheries Committee Annual Report(s) for 1970 and 1982.

13. British and Foreign State Papers, 27, p.983; C. 209. See UK Memorial, Annex A27 in ICJ Pleadings, The Minquiers and Ecrehos Case (United Kingdom/ France) Vol.1, 179-186. Leyden: A.W. Sijthoff, 1955.

14. Thomas Wemyss Fulton The Sovereignty of the Sea, pp.611-612. (Edinburgh and London: William Blackwood and Sons, 1911). Reprinted Millwood, New York: Kraus Reprint Co., 1976.

15. Dunn et al op.cit., p.5.

16. D.H.N. Johnson, 'The Minquiers and Ecrehos Case' The International and Comparative Law Quarterly, 3 (2) (1954), 189-216, at p.202.

17. D.P. O'Connell The International Law of the Sea Vol.1, p.516. Edited by I.A. Shearer, (Oxford: Clarendon Press, 1982).

18. Ibid, p.137. See also Fulton op.cit., p.612, 2, who states the area between the A-K line and the 3

mile limit is just over 100 square miles. All of the A-K line north of 49°3' is within the 3 mile limit, leaving about 23 square miles between the two limits in this area.

19. Johnson op.cit., p.203. All the figures in this chapter should be regarded as representational diagrams, rather than accurate maps.

20. Fulton op.cit., p.612.

21. Ibid, p.618.

22. See UK Memorial, Annex A 28 in ICJ Pleadings, The Minquiers and Ecrehos Case (United Kingdom/ France) Vol.1, 187-208. (Leyden: A.W. Sijthoff, 1955).

23. Johnson op.cit., p.202, n.40. cf. Fulton op.cit., p.619, who says the 1867 Convention was not ratified by France.

24. See Johnson op.cit., pp.192-193; and 'France-UK: Arbitration on the Delimitation of the Continental Shelf' International Legal Materials, XVIII (1978), 397-494, at pp.408-409 (para.6) (herafter ILM).

25. Johnson op.cit., p.205.

26. Ibid, p.202.

27. Lewis M. Alexander Offshore Geography of Northwestern Europe: The Political and Economic Problems of Delimitation and Control, p.128. London: John Murray, 1966). First published in the USA in 1963, (Chicago: Rand McNally and Co., 1963).

28. See Article 1 of the Special Agreement of December 29, 1950, submitting the case to the ICJ: Treaty Series No.103 (1951), Cmnd 8422.

29. Treaty Series No.2 (1929), Cmnd 3254.

30. See 'The Minquiers and Ecrehos Case, Judgment of November 17th, 1953' Report of Judgments, Advisory Opinions and Orders, 1953, (1954), 47-111 (hereafter ICJ Repts (1953)). Reprinted in International Law Reports, 20 (1954), 94-145. For commentaries see Johnson op.cit.; and Manley O. Hudson 'The Thirty-Second Year of the World Court'

American Journal of International Law, 48 (1) (1954), 1-22, at pp.6-12.

31. *ICJ Repts*, (1953), p.80.

32. *Ibid*, p.83.

33. The buoys are of no practical utility for Jersey, but only for vessels using the French port of St Malo. However, because the Minquiers are Jersey's sovereign territory, Jersey is forced to incur the costs of their maintenance. See Alexander *op.cit.*, p.129, n.1.

34. The use of circles to describe territorial sea limits may be unique.

35. France No.2 (1964), Cmnd 2363. See also Treaty Series No.54 (1968), Cmnd 3690.

36. This would have been impossible to regulate, as a Frenchman could hardly prohibit lobsters from entering his pots, whilst permitting crabs!

37. *Guernsey Evening Press and Star*, September 9, 1978.

38. *Ibid*, April 9, 1980. In 1985, a French fisherman admitted fishing illegally within Guernsey waters, less than 6 miles south of Sark. Guernsey claims an exclusive fishing limit of 6 miles, but under the terms of the 1964 European Fisheries Convention, permits habitual fishing rights to be exercised for certain types of fish in areas of its 6 to 12 mile belt. One of these areas is delimited by the L'Etac de Sark line. See *ibid*, July 5, 1985.

39. The exclusive French fishing rights are confined to the 'special regime' area, ie within the 1839/1928 A-K line. France's territorial sea was extended from 3 to 12 nautical miles by Law 71-1060, on December 24, 1971: 'National Claims to Maritime Jurisdictions', R.W. Smith (ed.), *Limits in the Seas* No.36 (5th Revision), p.64. United States Department of State, Bureau of Intelligence and Research, Office of the Geographer, 1985. Its straight baselines were proclaimed by Decree on October 19, 1967: see *ibid*. No.37 (1972).

40. This view was apparently held, although it is hard to see how legal agreements, such as exist,

could be easily disposed of.

41. Private correspondence with Capt. F.W. Jeune (Assistant Harbourmaster of Jersey, retd), August 28, 1986.

42. The government of Jersey.

43. The 1977 Court of Arbitration noted the UK had claimed the right to extend its territorial sea to 12 miles: 'France-UK: Arbitration on the Delimitation of the Continental Shelf' ILM, XVIII (1978), pp.397-494, at p.410 (para. 14).

44. Ibid.

45. Ibid, p.410 (para. 14).

46. Ibid.

47. Ibid, (para. 17).

48. Ibid, p.410 (para. 16).

49. Ibid, (para. 17).

50. Ibid, (para. 18).

51. Ibid, p.411 (para. 22).

52. The UK and France are in agreement that, in principle, the boundary line between the Channel Islands and France should be a median line: see ibid, pp.410 (para. 15), 411 (para. 22).

53. Ibid, p.411 (para. 18).

54. Ibid, p.410 (para. 15). The UK also contests France's straight baseline across the Anse de Vauville.

55. Supra, note 52.

56. Guernsey Evening Press and Star, April 1, 1982.

57. Supra, note 41.

58. Alexander op.cit., pp.129-130.

59. It has been argued, that it was in the UK's best interests that the Island receive as little

territory as possible out of the continental shelf arbitration with France: see <u>Guernsey Evening Press and Star</u>, April 1, 1982. The UK might well give resolving the <u>mer commune</u> a low priority in its relations with France. If so, the Islands' best interests will again be secondary to the UK's.

60. If this case were to go to arbitration, France might well argue that, on the basis of the agreements perpetuating the <u>mer commune</u> arrangement, the waters concerned are 'historic waters' belonging to both France and Jersey.

61. The likelihood of hydrocarbons being found in these narrow waters is regarded by experts as geologically remote.

62. Lewis M. Alexander 'Offshore Claims and Fisheries in North-West Europe' <u>The Yearbook of World Affairs</u>, XIV (1960), pp.236-260, at p.249. As late as 1895 there were nearly 3,000 fishermen in the Bay of Granville.

63. Dunn et al <u>op.cit.</u>, p.122.

64. Derwent Whittlesey <u>The Earth and the State</u>, p.565, (New York: Henry Holt, 1944). Cited by Alexander (1966) <u>op.cit.</u>, 136-137.

CHAPTER 8

EUROPEAN, NATIONAL AND REGIONAL CONCEPTS OF FISHING
LIMITS IN THE EUROPEAN COMMUNITY

Mark Wise

Conflict over the European Community's Common
Fisheries Policy (CFP) is usually portrayed in
national terms. Complex clashes of interest
involving diverse groups operating at different
geographical scales are popularly perceived as crude
contests between 'the British', 'the French', or
whoever. The political structure of the European
Community (EC) exacerbates this tendency in that
member-state governments remain the most powerful
decision-making element within it. The different
national fishery ministers who meet to decide policy
in the EC's Council of Ministers inevitably define
their individual objectives in the knowledge that
they are primarily dependent upon their particular
national parliaments, electorates and interest
groups, rather than European bodies. (1)
However, discord over access to fishing grounds
within the EC is not a simple inter-national
struggle among separate member states. The Community
is governed by supra-national institutions including
the European Commission which has the task of
adopting a 'European' perspective and formulating
policy proposals accordingly. Consequently, it has
striven to develop systems of access to fishing
grounds which break free of the 'national' ways of
thought underlying most maritime boundaries (in this
chapter the word 'national' will be used to refer to
activity at the member-state scale, while 'European'
refers to the EC level). Sometimes, the Commission's
promotion of a 'European' view amounts to little
more than an attempt to find a compromise among
opposing national interests, but at others it has
represented a fundamental attack on the national
discriminations involved in member-state practices.
As a result, a tension between 'European' and
'national' concepts of fishery management has been a

persistent element in the development of the CFP. This has helped to generate ideas about access to sea-fish resources which relate to geographical scales other than the Community as a whole or its individual states. For example, so-called 'regional' systems have proved attractive, not only in that they apply to spatial entities appropriate for tackling the management problems at hand, but because they conform to a Community principle which aims to eliminate national discrimination among member states (the term 'regional' is used here to refer to geographical scales other than 'European' or 'national' as defined above). This chapter analyses the complex interplay between these 'European', 'national' and 'regional' concepts of sea fishery limits in the evolution of the CFP. It will show how regional ideas have had some impact on the issue of 'who can fish what and where' in EC waters, although the reality of restrictive national fishery zones in a Community supposedly dedicated to the eradication of such discriminations is still very evident.

ORIGINAL PROPOSALS: COMMUNITY FREE OF NATIONAL DISCRIMINATION

The original proposals for a CFP required that 'Community' fishermen must have access to and use of fishing grounds in the maritime waters (2) coming under the sovereignty or within the jurisdiction of member states. This did not mean an open-access 'free-for-all' in Community waters, but any regulations within the fishing zones of member states would have to be applied equally to all EC fishermen regardless of nationality. These proposals were based on the principle laid down in Article 7 of the Treaty of Rome that '... any discrimination on grounds of nationality shall be prohibited' among member states. (3) This so-called 'equal-access' provision was included in the first CFP agreement adopted by the original 'Six' EC states in June 1970 just prior to the opening of enlargement negotiations with Britain, Denmark, Ireland and Norway. (4) This timing led many in the applicant states to suspect - not without some reason - that this elimination of national discrimination from fishing access arrangements had rather less to do with 'European' ideals than opportunism of a nationalistic nature. All the original EC states faced increasing restrictions on access to fishing grounds and had an obvious interest in supporting a Community principle which gave their fishermen the prospect of freer entry into the progressively wider exclusive national fishing zones around 'fish-rich' countries as Norway, the Faroes, Greenland, Britain and Ireland. (5)

NORWAY'S REGIONAL CONCEPTS REJECTED DURING THE COMMUNITY'S FIRST ENLARGEMENT

When the applicant states began to negotiate about fisheries in June 1971, they found that the CFP, with its equality of access provisions, was presented to them as part of established EC law which had to be accepted. This posed difficulties for all of them, but only the Norwegians rejected the CFP's access provisions without equivocation, insisting on retention of what was then their 12-mile national fishing limit and reserving the right to extend it in the future. This national stand was thought essential to protect Norwegian fishermen living in the harsh environment of the country's coastal regions. (6) In response to the argument that this would involve national discrimination, the Norwegians stressed that their protective fishery policies formed part of a 'social philosophy' which differentiated between Norwegian citizens let alone those from other countries. For example, only Norwegians actively engaged in fisheries were normally allowed to own fishing vessels. Such discriminatory legislation was designed '... to prevent capital interests from outside fishing communities and outside the fishing districts from obtaining a dominant position in the fishing industry'.(7) Such domination would, in Norway's view, lead to the economic elimination of independent, small-scale fishermen and a subsequent depopulation of remote coastal regions.

However, the Norwegians insisted that a solution existed which reconciled both their national interests and 'European' principles. They proposed a compromise of a regional character based on the 'right of establishment' provisions incorporated in Rome Treaty. (8) Citizens of other EC states would be permitted to fish within Norwegian fishing limits on condition that they were resident in Norway and had registered their vessels in that country. In other words '... those who want to exploit these limited resources should share the lot of those who live there...' to ensure that this exploitation is '... undertaken on equal terms'. (9) Such a regional solution was also justified by Norway on the grounds that it alone would enable conservation measures to be effectively enforced within its fishing limit'... the long nature of the Norwegian coastline makes it impossible to control fishing effort unless the fishing fleet which is allowed to fish here is attached to the coast by establishment there'. (10) However, despite the ingenuity of this regional access scheme based on Community law, it was rejected by the original 'Six'. After some extremely intricate negotiations, an agreement was reached and incorporated into the Treaty of Accession. (11) This uneasy accord was based on a temporary compromise between 'European' and 'national' concepts of fishery limits, with faint traces of the 'regional' ideas outlined above. Although the principle of a Community access system without national discrimination was maintained, the original EC states were forced to accept that its full implementation would have to be deferred for at least 10 years. During that period member states would be able to preserve a 6-mile national zone from which the citizens of other Community countries could be excluded. In certain regions deemed to be particularly dependent on fishing, this limit could be extended out to 12 miles. This provision could be perceived as discriminating in favour of particular groups in specific areas rather than a whole national population of fishermen. Thus, belief in progress towards a genuinely 'European' system of fishing rights could be somewhat tenuously preserved. Indeed, even the more obviously national 6-mile limit maintained around the whole of a member state's coastline was conceded in language which gave some succour to those struggling to keep Community concepts alive; Article 100 of the Treaty of Accession stated with deliberate ambiguity that '... the member states of the Community are authorized ... to restrict fishing in waters under their sovereignty ... situated within a limit of six nautical miles ... to vessels which fish traditionally in those waters and which operate from ports in that geographical coastal area'. (12)

So the applicant states won concessions en-abling them to perpetuate national discrimination in controlling access to sea-fish resources. But in the legal foundations of the CFP, restrictions on fishing access were still largely couched in terms of geographical areas not necessarily concordant with national territories. This was doubtless one reason why the Norwegian electorate eventually rejected the terms negotiated by its government for entry into the EC, with the result that Europe's most important fishing state remains beyond the realms of the CFP. Norwegian fishermen wanted unambiguous guarantees that entry into 'their' waters by fishermen from EC countries would be indefinitely prevented by a clear national fishing limit, undiluted by subtle schemes based on 'European' and/or 'regional' concepts.

EQUALITY OF ACCESS OR EXCLUSIVE NATIONAL FISHING ZONES WITHIN THE 200-MILE LIMITS OF COMMUNITY STATES?

During the 1970s the move towards 200-mile/ median line Exclusive Economic Zones proved un-stoppable. Consequently, the member states of the EC acted in concert on January 1st 1977 to extend their fishing zones. Together, these separate national limits enclosed a vast area of sea which became known unofficially as the 'Europond' (Figure 8.1). This radical change in the political geography of the marine space encircling the EC required reform of the CFP. However, it was not until January 1983 that this was achieved. In the lengthy conflict over who should fish how much in what areas, European, national and regional concepts of fishing access again interweaved with one another in complicated combinations, leading eventually to a rather 'hybrid' system of control.

In simple terms, the conflict could be reduced to an argument between the continental states on one hand and the countries of the British Isles on the other. Whereas the former wanted a 'Community' system essentially free of national limitations on access to fishing grounds, Britain and Ireland were determined to achieve a substantial measure of preference for their fishermen through the establishment of national fishing zones closed to their Common Market partners.

With their limited coastal resources, traditional patterns of fishing around the British Isles and loss of distant-water opportunities, the continental states were bound to favour a 'European' approach to access rights in the so-called 'Europond'; for them, national interest and Community principle coincided. They could continue to point to the national non-discrimination provisions embedded in the existing CFP regulations

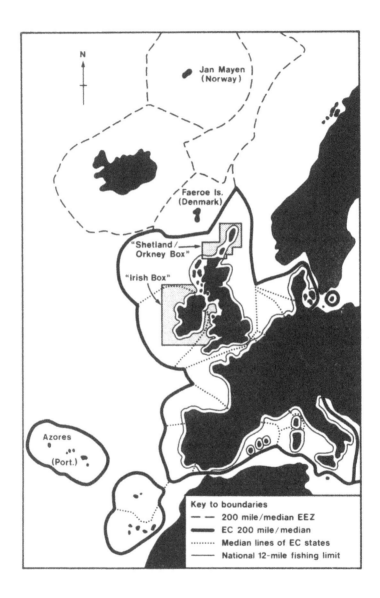

Figure 8.1. Various fishing limits within the European community (not to scale)

and argue that the extension of fishing limits to 200 miles made no difference to this basic Community principle. The economic argument that the free movement of fish products within the Common Market must be counterbalanced by fishing access arrangements which also do not differentiate according to nationality was persistently made. (13) Moreover, the Irish in particular were told that it was unreasonable to reap the rewards of national non-discrimination from Community policies dealing with agriculture and poorer regions, and then ignore it when fishing rights were involved. The continental states also lined up behind the Commission's argument that effective conservation of resources in the 'Europond' required Community controls rather than national fishing limits. The biogeographical reality that fish are mobile and do not respect the arbitrary geometry of national fishing zones clearly reinforced this point. In repudiating national notions of fishery management in EC waters, it was often stressed that the Community had the political and legal capacity to enforce effective conservation measures of a 'European' character. (14)

So, Britain and Ireland apart, the EC member states largely endorsed the Commission's arguments that exclusive national fishing zones of an extensive nature should not be used as a means of fishery resource allocation and conservation among Community states. But what sort of 'European' system compatible with Community principles could take their place? In very general terms the following system was advocated. (15) First, the Community would establish total allowable catches (TACs) for all the important fish stocks. Then it would divide them into quotas according to historic fishing patterns and the 'vital needs' of regions 'particularly dependent' upon fishing (ie: Ireland, Greenland and northern Britain). Once these conservational TACs and allocational quotas had been decided, they would be enforced, along with other EC fishery measures, as part of Community law by both national and European authorities.

Such proposals found little favour in Britain and Ireland. The UK had lost its final 'cod war' with Iceland in 1975 with the result that British distant-water fishermen, long the champions of open international seas, swung round to support their inshore compatriots pushing for a wide national fishing limit which would exclude operators from other EC countries as well as those from further afield. (16) For example, by January 1976, British

United Trawlers Limited of Hull was calling for a
100-mile/median line fishing zone - within the 200-
mile EEZ - which would be for the exclusive use of
UK nationals. (17)

In defending their demands for extensive,
exclusive national fishing limits within the
'Europond', the British and Irish mustered arguments
which have become familiar as states have pushed
their way out to 200-mile maritime boundaries.
Attempts to control rampant overfishing through
international organisations like the North East
Atlantic Fisheries Commission (NEAFC) had proved
largely futile. Therefore, the extension of national
sovereignty over sea space was presented as the only
effective means of controlling fishing effort and
protecting fishery resources. Crude as they might be
in eliminating foreign fishermen in an arbitrary
manner, national fishing limits had, it was argued,
the virtue of being clear, enforceable and able to
make substantial reductions in excess fishing
capacity exploiting depleted stocks. (18) More
sophisticated international schemes of allocation
and conservation of the sort envisaged by the
European Commission would be subject to the
indecision and lax compromises inherent in international organisations. (19) Where international
co-operation was required to protect and share out
mobile fish stocks, this could best be achieved
through intergovernmental accords at a scale most
appropriate for the problem at hand; for example,
management of fish migrating between the UK and
Norwegian sectors of the North Sea could be most
efficiently handled through simple bilateral
arrangements agreed by the two countries directly
concerned without being embroiled in complex, multi-
governmental EC procedures.

Another line of argument used by the British
and Irish in pressing for a system of fisheries
management based on exclusive national zones related
to regional development. Analyses of socio-economic
welfare throughout the European Community always
reveal that Ireland along with northern and western
Britain are among the poorest regions. (20) Surely
the aims of the EC's regional policy would require
that industries - such as fishing - with potential
for growth in these unfavoured areas be shielded
from the full force of common market competition?
The Irish in particular felt that their small, but
growing, fishing industry would be overwhelmed by
large, modern fleets of other EC states unless a
wide exclusive fishing zone was established around

its shores in place of a more open 'European'
system. (21)

The British and Irish also pointed to the fact
that their fishing zones made up around 60 per cent
and 10 per cent respectively of the 'Europond' in
the North-East Atlantic in justifying their claims
for essentially national systems of fisheries
management (Figure 8.1). It was easy to reason that
the UK and Ireland should be allocated shares of the
total EC catch in proportion to these zones and that
exclusive national fishing limits would make sure
that they got them. This was backed by the feeling
that the CFP was unfairly requiring the British and
Irish to share 'their' national fishery resources
with other EC states while no such common European
ownership was demanded elsewhere. (22) This
sentiment ignored the complexity of the real
situation. The continental states were demanding
equality of access to marine resources regardless of
nationality rather than common ownership of them.
Furthermore, although resources like agricultural
land remain 'national' assets, the rights of establishment laid down in the Treaty of Rome allow
citizens of one member state to settle in another
and exploit them (see discussion of Norway's
regional proposals above).

REGIONAL FISHING PLANS PROPOSED TO RECONCILE
EUROPEAN AND NATIONAL VIEWPOINTS

In the efforts to resolve the conflict between
these opposing Community and national approaches to
fisheries control, the idea of regional fishing
plans was promoted. These plans were designed to
meet British and Irish demands for restricted access
to waters around their shores, but in a way which
did not violate the Community's national non-
discrimination principle. Such plans would 'promote
rational exploitation of the biological resources
bearing in mind the social and economic needs of
certain categories of fishermen in specific regions
of the Community' and 'assure, in regard to these
regions, the enjoyment of the natural geographic
advantage in catch possibilities within a few hours
steaming time from home ports'. They would '... not
discriminate as between fishermen of the member
states of the Community or affect their right of
access', but they would '... take into account that
vessels which, due to their limited range of
operation, can only exercise their activities close

to the coast, should have priority in the coastal areas'. Furthermore, '... the activity of other categories of vessels must be harmoniously introduced into the global fishing activity of all the vessels operating in the area and, in particular, undue concentration of long-range vessels in areas closest to the coast should be prevented. (23) In other words, these plans would not differentiate between Community fishermen according to citizenship, but would discriminate in favour of non-national groups of fishermen in specific regions where conservation of fish stocks and jobs so justified. It was this concept of fisheries control which contributed to the Irish government's decision to abandon the demand for wide exclusive national fishing limits within the 'Europond' in early 1978, although the British government remained sceptical of this regionally sensitive concept of fishing rights.

COMPROMISE BETWEEN EUROPEAN, NATIONAL AND REGIONAL CONCEPTS IN THE CFP AGREEMENT OF JANUARY 1983

After six year's bargaining of Byzantine intricacy, a new CFP was agreed in January 1983 based on compromises struck between the various approaches outlined above. (24) Those seeking a 'European' system of a fisheries management could be satisfied with some elements in the new accord. For example, it is the Community which fixes annual TACs for the main stocks in the 'Europond'. These allowable catches, supported by a whole range of other conservation measures decided by the Community, are divided into quotas according to the following criteria: traditional fishing patterns; the needs of regions especially dependent on fishing; and, least important, the loss of catch potential in third-country waters. In an effort to avoid recurring national conflict over the share-out of TACs, the 1982 distribution of quotas has been laid down as a 'reference allocation' to determine future divisions of sea-fish stocks available to the EC. On the face of it, there is nothing in these arrangements to offend Community principles. However, the dominant position of member state institutions in the EC's political structure ensures that national discrimination remains ingrained in the new system. For example, quotas are still allocated country by country, although these national shares (Britain obtains about 36 per cent of the seven major species) have to be caught in particular fishing zones specified by the <u>Community</u>. Regarding access to fishing zones, 'European' idealists can argue that the <u>principle</u> of eliminating national discrimination has not been abandoned and point to the fact that Community waters have not been carved up into the wide exclusive national fishing zones originally demanded by the British and Irish. Furthermore, they can cite the 'Shetland/Orkney box' as a fruit of the <u>regional fishing plan</u> concept (Figure 8.1). Within this non-national zone, a degree of preferential access has been accorded to <u>local</u> (n.b. not <u>national</u>) fishermen by discriminating against vessels over 26 metres in length catching demersal species. This <u>non-national</u> discrimination permits the smaller local boats to fish in the 'box' without a licence, whereas operators of larger vessels from further afield have to obtain one if they wish to exploit this restricted zone which has clearly been established to protect jobs and fish stocks in a peripheral region. Such an arrangement appears impeccably 'communautaire' in character, but the licences available to larger vessels are allocated on a <u>national</u> basis: 62 to Britain, 52 to France, 12 to West Germany, and two to Belgium. Once again, the hybrid fishing access system emerging from the cross-fertilisation of European, national and regional ideas within the EC is illustrated.

Beyond regional plans, national notions of fishing limits are by no means dead in the so-called 'Europond'. The temporary exceptions to the equality of access principle permitted in the 1972 Treaty of Accession (see above) were extended in 1983 in both spatial and temporal terms. First, the six-mile national fishing belt, far from receding when the 10-year transitional period came to an end in December 1982, has been pushed out to 12 miles around the <u>whole</u> EC coastline, although the traditional rights of other member states in these extended national areas must be respected. These further exceptions to the concept of a 'European Community fisherman' will last until 2002 when the situation will be reviewed.

EUROPEAN PRINCIPLES, NATIONAL INTERESTS AND REGIONAL ACCESS RESTRICTIONS AS SPAIN ENTERS THE COMMUNITY

As the Iberian countries negotiated their way towards EC membership in the early 1980s, lurid talk of a new 'Spanish Armada' invading northern European waters became common quayside currency, suggesting that national discrimination, overt and covert, is likely to be part of Community fishery life for the foreseeable future. Integration of Portugal's essentially inshore industry posed no great difficulties, but absorption of Spain's enormous fishing fleet was a different matter. In 1982, there were some 106,600 Spanish fishermen operating some 17,500 vessels with a combined tonnage of some 738,500 GRT; at the same time in the EC 'Ten' there were around 53,850 vessels totalling 1,116,000 GRT, used by some 160,000 fishermen. (25) Although Spain was taking only about 9 per cent of its catch in EC waters at this time, Spanish negotiators were pointing to the national non-discrimination principles of the CFP, and the fact that Spain would represent a highly lucrative outlet for fish exports within the common market, in order to press for a larger share once in the Community. This provoked unambiguously national reactions from the existing member states who were determined to keep the Spanish out of 'their' waters. These fears were fuelled by the knowledge that the Spaniards were urgently seeking new fishing opportunities following their exclusion from 200-mile EEZs around the world; the EC itself had substantially reduced Spanish activities in the existing 'Europond' (by 1982 the Community was granting Spain a mere 106 licences to operate in waters where the Spaniards claimed to have had some 600 boats fishing in the early 1970s).

The enlargement negotiations thus reverberated with familiar echoes of earlier debates, although changing patterns of national interest sometimes led the actors to amend their lines; for example, the French, firing shots across the bows of Spanish vessels in the Bay of Biscay, were no longer defending the Community's equal-access principle with the vigour they had displayed when entry of their fishermen to the waters around the British Isles was at stake (see above). Finally, the inevitable compromise falling somewhere between Community ideals and national realities was constructed. (26) Spain, like Portugal, was incorporated into the CFP system immediately following its entry into the EC in January 1986. This meant that it was accorded quota shares of certain Community TACs like other member states; for example, a hake quota of 18,000 tonnes was agreed, to be caught under normal CFP conditions where Spain is required to comply with the same common rules applying to all EC fishermen regardless of nationality.

But such acceptance of the Spanish into the 'European' fold was tempered by a whole range of access restrictions which, although sometimes having a 'regional' aspect, were clearly aimed at Spanish nationals alone. First, the number of Spanish vessels able to operate in EC waters has been limited to 300 listed vessels of which only 150 are allowed to fish at any one time. These boats can only operate in specific regions of the 'Europond' where the Spaniards can legitimately claim some traditional rights: 57 vessels in the Bay of Biscay; 23 to the north of Ireland and west of Scotland; and 70 in the English Channel and waters around Ireland. Moreover, access to the large 'Irish box' (Figure 8.1) has been denied to Spanish fishermen until 1995. The North Sea region has also been closed to them by a similar national discrimination. During the 10-year transition period, Community funds will be used in an effort to reduce surplus capacity in the Spanish fleet so that equality of access to these zones by the mid-1990s will not seem such a threat to the developing Irish fishing industry and those dependent on the crowded North Sea arena.

CONCLUSION

This chapter has revealed the <u>regional</u> dimension in EC fishing limit disputes which are usually presented in crude <u>national</u> and <u>Community</u> terms. In fact, concepts based on all three geographical scales have intertwined with each other throughout the development of the CFP. The regional ideas provoke thought about fishing access arrangements which may, in some circumstances, be more appropriate than those associated with the 'European' and national scales. In reality, the fishing industry within the various member states of the European Community breaks down into a myriad pattern of competing regionally-based communities; for example, how meaningful is it to talk of 'British fishermen' as a distinct national entity given the hostility that often exists between such different groups as Cornish handliners and Scottish purse-seiners? Regional ideas, relating to areas other than the somewhat arbitrary ones associated with the EC as a whole and its individual member states, offer a way of recognising such diversity in the making of maritime boundary systems designed to conserve and allocate the fishery resources of the Community. It is possible to speculate that Norway would now be operating within the CFP if such regional concepts had been more acceptable at the time it was seeking to enter the Common Market. If that were the case, the Community would today be developing fishery policy within the spatial framework of a more genuine 'Europond' extending from the Azores to Spitsbergen.

NOTES

1. See M. Wise, <u>The Common Fisheries Policy of the European Community,</u> (London, Methuen, 1984) for elaboration of this point.

2. 'Politique commune de la pêche: proposition de règlement du Conseil portant établissement d'une politique commune des structures dans le secteur de la pecherie', <u>Journal Officiel des Communautes Européennes</u>, 13/9/68, C 19, 1-19.

3. 'Treaty establishing the European Economic Community, 25 March 1957, Rome', <u>Treaties Establishing the European Communities</u> (EEC, Luxembourg, 1973).

4. 'Council Regulation (EEC) 2142/70 of 20 October 1970 laying down a common structural policy for the fishing industry', <u>Official Journal of the European Communities</u>, 27/10/70, 14, L 236, 1-4.

5. Wise, <u>op.cit.</u>, Ch.4.

6. Norwegian Government Declaration, (8/6/71), (Commission of the European Communities, internal documentation, Brussels).

7. <u>Ibid</u>.

8. 'Treaty establishing the European Economic Community, 25 March 1957', <u>op.cit.</u>, Articles 52-58.

9. Norwegian Government Declaration, (8/6/71), <u>op.cit</u>.

10. Norwegian Statement at EEC Ministerial Meeting, (9/11/71), (Commission of the European Communities, internal documentation, Brussels).

11. 'Act concerning the Conditions of Accession and the Adjustments of the Treaties', Articles 100-103, Treaties Establishing the European Communities, (Luxembourg, 1973).

12. Ibid, Article 100.

13. J. Regnier, 'The real meaning of Community', in Fisheries of the European Community (White Fish Authority, Edinburgh, 1977).

14. E. Gallagher, 'The Community's Fisheries Policy' in Fisheries of the European Community, (White Fish Authority, Edinburgh, 1977), 2.

15. 'Future external fisheries policy: an internal fisheries system', COM(76)500 final, 23/9/76. (Commission of the European Communities, Brussels, 1976).

16. A. Underdal, The Politics of International Fisheries Management: the Case of the Northeast Atlantic, (Universitetsforlaget, Oslo, 1980).

17. Proposals for a UK Fishery Policy, (British United Trawlers Limited, Hull, 1976).

18. N. Buchanan, and D. Steel. 'Meaningful effort limitation: the British case' in Fisheries of the European Community, op.cit. 7.

19. I.C. Wood, 'Towards a Common Fisheries Policy, a British Viewpoint', Eurofish Report, (7 June 1978).

20. 'The regions of Europe: second periodic report on the social and economic situation and development of the regions of the Community', COM(84)40, 4/4/84, (Commission of the European Communites, Brussels, 1984).

21. A. Heskin, 'A profile of the Irish fishing industry', in Fisheries of the European Community, op.cit., 14-16.

22. I.C. Wood, 1978, op.cit.

23. 'Draft resolution of the Council concerning the introduction of fishing plans', COM(78)39, 30/1/78 (Commission of the European Communities, Brussels, 1978).

24. 'Council Regulation (EEC) 170/83 of 25 January 1983 establishing a Community system for the conservation and management of fishery resources', Official Journal of the European Communities, 27/1/83, 26/L 24, 1-13; and 'Council Regulation (EEC) 172/83 of 25 January 1983 fixing for certain fish stocks and groups of fish stocks occurring in the Community's fishing zone, total allowable catches for 1982, the share of these catches available to the Community, the allocation of that share between the member states and the conditions under which the total allowance catches may be fished', Official Journal of the European Communities, ibid, 30-67.

25. The European Community's Fishery Policy (Commission of the European Communities, 1985, Luxembourg).

26. 'Act concerning the conditions of accession of the Kingdom of Spain and the Portuguese Republic and the adjustments of the Treaties', Official Journal of the European Communities, 15/11/85, 28, L 302, 69-74.

CHAPTER 9

OFFSHORE JURISDICTIONAL CLAIMS OF THE REPUBLIC OF IRELAND

Proinnsias Breathnach

INTRODUCTION (1)

The evolution of the offshore jurisdictional claims of the Republic of Ireland beyond the limits of the territorial sea coincides with the extension of hydrocarbon exploitation into deeper water areas[2] and corresponds with the proposed revision of international sea law through the Third United Nations Conference on the Law of the Sea (UNCLOS). (2) This chapter both outlines the progressive sequence of Irish offshore jurisdictional claims (3) and considers the implications of counterclaims by adjacent states to areas also claimed by Ireland, paying particular attention to the question of boundary demarcation between Ireland and the United Kingdom.

THE 200 METRE ISOBATH

Ireland's interest in the offshore area beyond the traditional three mile territorial sea was first stimulated by the discovery of substantial hydrocarbon resources in the North Sea in the 1960s. Subsequent surveys of the offshore areas around Ireland indicated a number of sedimentary basins with hydrocarbon potential (Figure 9.1) and the legal framework for claiming jurisdiction over such areas was created via the 1968 Continental Shelf Act.

The first three designations were made in 1968 (Figure 9.1), and remained within the limit of the 200 metre isobath, identified by the 1958 United Nations Convention on the Continental shelf as the normal limit of a coastal state's offshore seabed resource jurisdiction. The 1968 designations also

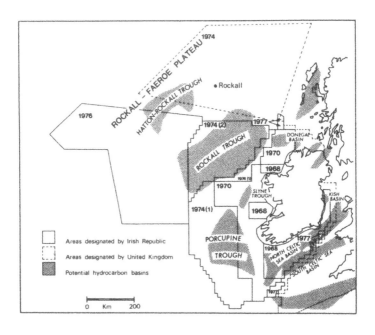

Figure 9.1. Chronology of Irish seabed designations

fell short of where it was thought the boundary line
with Britain would lie to the east and southeast.
Following these designations, all three areas were
exclusively licensed to the Marathon Oil company.
These include the North Celtic Sea Basin and part of
the South Celtic Sea Basin (Figure 9.1).

   In 1970, the Irish government extended its
jurisdiction off the west coast to an area roughly
coincident with the 200 metre isobath (Figures 9.1
and 9.2), thus incorporating the Slyne Trough and
Donegal Basin within its sphere of control.

THE CONTINENTAL SHELF

   While the 200 metre isobath had originally been
used (in 1945) by the United States to define its

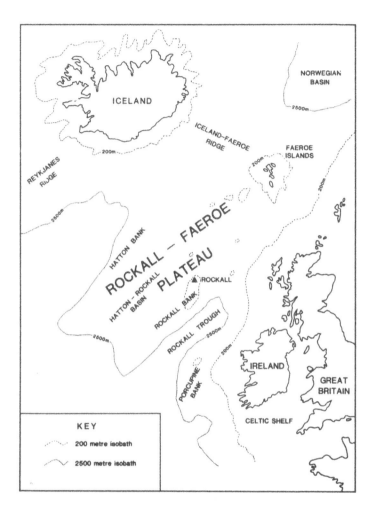

Figure 9.2. Bathymetry and topography of the Northeast Atlantic

continental shelf (4), in the Irish case the shelf extends well beyond this limit over a large area. With offshore hydrocarbon exploration technology developing rapidly, in line with the extension of exploration into the deeper areas of the North Sea and receiving a further boost from the 1973/4 oil crisis, the possibility now existed for exploration throughout the Irish continental shelf, including the attractive Porcupine Trough, most of which lies beyond the 200 metre isobath limit (Figure 8.1).

Thus, early in 1974, the Irish government made a further designation extending approximately to the base of the continental slope (Figure 9.1). Such a claim was allowed for in the 1958 UN Convention via the clause permitting resource exploitation beyond the 200 metre isobath where this was feasible (the so-called 'exploitability criterion'). (5) A considerable amount of exploration has since been carried out in the Porcupine Trough, and some (so far) uncommercial discoveries have been made.

THE 200 NAUTICAL MILE LIMIT

When UNCLOS convened for its first full working session in 1974, initial attention was focussed on the idea of a minimum 200 nautical mile (nm) zone (representing the average width of the continental shelf in the Atlantic) wherein the coastal state would have exclusive control of all natural resources. This limit was designed, in part, to compensate those states which, for geological reasons, have a restricted natural prolongation of their landmasses into the adjoining sea. The crystallisation of this idea spurred the Irish government to make a second designation in late 1974 extending jurisdiction to the 200 nm limit (Figure 9.1).

This designation incorporated most of the Rockall Trough, lying between the Irish continental shelf and the submarine Rockall-Faeroe Plateau to the northwest but with the exception of Rockall (Figures 9.1 and 9.2). This latest claim may be seen, in part, as a response to the designation by the British government earlier in the same year of a large tract of seabed territory surrounding Rockall (Figure 9.1). The Irish claim, coming within 100 nm of Rockall, thus represented a pre-emption of any attempt by Britain to declare a 200 nm limit around Rockall, (6) (see below).

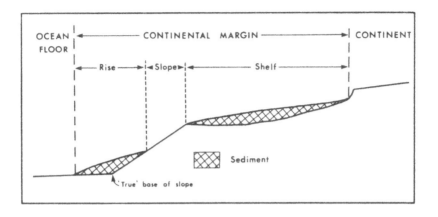

Figure 9.3. Model of the continental margin

THE CONTINENTAL MARGIN

As UNCLOS progressed, the concept of the continental <u>margin</u>, comprising the shelf, slope <u>and</u> rise (Figure 9.3), came increasingly into focus. (7) It was argued that the rise consisted of sediments deriving originally from the shelf and therefore should also be regarded as part of the natural prolongation of the adjoining landmass, thereby coming under the jurisdiction of the coastal state. Using this 'natural prolongation' argument, the Irish government in 1976 extended its area of jurisdiction to the western edge of the Rockall-Faeroe Plateau, including the southern portion of the Hatton-Rockall Trough (Figure 9.1). At its furthest point, the area thereby claimed reaches more than 500 nm from the Irish coast. This claim brought the total area designated by the Irish government to over 500,000 sq. km, or over seven times the area of Ireland's landmass.

The basis of Ireland's 1976 designation has been called into question, (8) on the grounds that the Rockall-Faeroe Plateau is geologically distinct from the Irish continental shelf, and could more

accurately be regarded as the natural prolongation
of the Faeroe Islands, some 750 nm to the north
(Figure 9.2). In addition, it has been suggested
that the claim may contradict Article 76 of the
UNCLOS Convention which finally emerged in 1982, (9)
which restricts continental shelf claims to a
maximum of 350 nm from the coast, or 100 nm from the
2,500 metre isobath, (10) both of which, it has been
argued, fall well short of the outer limit of the
area claimed by the Irish government (Figure 9.2).
These latter restrictions, it may be noted, had not
been included in the UNCLOS negotiating text at the
time the Irish claim was made. In any case, there is
some doubt whether there is a true geological
connection between the Faeroe Islands and the
Rockall-Faeroe Plateau. (11) The Irish position (12)
is that the latter Plateau is, both geologically and
legally, part of the Northwest European continental
Shelf, and comes partially within Ireland's own continental shelf, as legally defined by UNCLOS (13);
accordingly, the 2,500 metre isobath limit only
applies to the western margin of the Plateau.

THE DANISH AND ICELANDIC COUNTERCLAIMS

For nine years Ireland's 1976 claim remained
undisputed. However, the situation changed
dramatically in 1985 when both Denmark and Iceland,
in close succession, made claims to vast tracts of
territory in the region in question (Figure 9.4).
The Danish claim (on behalf of the Faeroe
Islands), made on May 7, 1985, incorporates that
part of what they referred to as the 'Faeroe-Rockall
micro continent' lying beyond the British and Irish
200 nm limits. The claim covers an area of 300,000
sq. km and stretches to over 1,000 nm in length at
its maximum extent. It includes the bulk of the area
designated by Ireland in 1976, and part of the area
around Rockall designated by Britain in 1974.
In a statement accompanying its claim, the
Danish government stated that its action was not an
unfriendly gesture towards Britain or Ireland, but
rather a statement of prior claim in advance of
negotiations between the interested parties concerning the delimitation of national jurisdictions
in the region in question. A failure to make a
designation could be interpreted as an acceptance of
British and Irish claims in the area. (14)
Three days later, on May 10 1985, Iceland made
an even more extensive claim to offshore territory

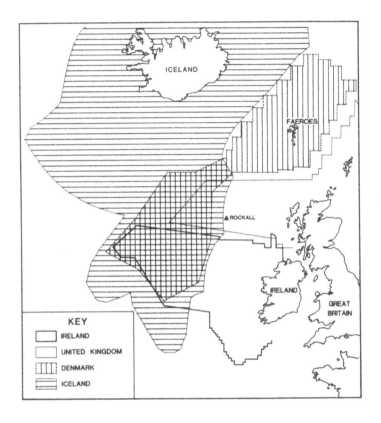

Figure 9.4. Offshore jurisdictional claims of Denmark, Iceland, Ireland and the United Kingdom in the Northeast Atlantic

(Figure 9.4). The eastern boundary of the Icelandic designation is determined, respectively, by the median line with the Faeroes and by the British and Irish 200 nm limits, while the western boundary is determined by the median line with Greenland. The boundary to the southwest is determined by the 350 nm limit specified for submarine ridges by Article 76 of the UNCLOS Convention, while the southern boundary (some 1,000 nm from Iceland at its furthest point) is determined by the line 60 nm beyond the foot of the continental slope also specified in the same article.

Iceland's claim would appear to be particularly flimsy. Not only does the southern boundary of the claim extend much further than the 100 nm beyond the 2,500 metre isobath allowed for in Article 76 of the UNCLOS Convention, but the Rockall-Faeroe Plateau, which consists of true continental shelf, is geologically distinct from Iceland, an island of volcanic origin lying atop the mid-Atlantic Reykjanes Submarine Ridge (Figure 9.2).

As with the Danes, the Icelandic claim was not aggressive, Article 5 of the Ministerial Order under which the claim was made states: 'Agreement between Iceland and the other countries concerned is to be sought on the definitive delimitation of the Continental Shelf area south of Iceland in accordance with the general rules of international law'. This is a similar form of words to that used in the Danish claim. Indeed, there have been suggestions that Iceland has been seeking the co-operation of Denmark in forming a common front in any negotiations with Ireland and Britain. (15) Any urgency with such negotiations has, at the time of writing in 1986, greatly receded in the light of the collapse of international oil prices, given the prohibitive water depths and climatic conditions generally prevailing in the region in dispute.

THE IRELAND-UNITED KINGDOM BOUNDARY

Despite a considerable amount of exploratory drilling, and some significant hydrocarbon finds in the Celtic and Irish Seas, and the awarding by Britain of exploration licences in disputed waters in the Rockall region, progress towards the delimitation of an agreed boundary between Ireland and the United Kingdom has been extremely slow.

Before 1976, both sides had been careful to only designate areas which had not been previously designated by the other. However, Britain's designation of a 135,000 sq. km area around Rockall in 1974 brought forth a protest from the Irish government, which maintained that the designation included areas which 'as a matter of international law fall within Irish jurisdiction' and 'which are closer to the Irish than to the British coast'. (16) A British government spokesman retorted that the area designated was 'entirely a prolongation of the landmass of Scotland and is an area over which the United Kingdom has sovereign rights in international law'. (17) The spokesman added that the claim was not based on British ownership of Rockall.

Rockall was originally annexed by Britain in 1955 as a security measure to prevent the possible use of the rock for surveillance of the missile base then being developed in the Outer Hebrides. Rockall (which is merely 22 metres high with a circumference of 25 metres at the base) was incorporated into Inverness-shire by Act of Parliament in 1972. (18) Although Rockall is closer to the Irish than the Scottish mainland, Ireland has never challenged (nor accepted, for that matter) Britain's ownership thereof, but has consistently maintained that the rock has no status in relation to the delimitation of offshore resource jurisdiction. (19)

Despite Ireland's position on Rockall's status, in 1976 Britain declared new fisheries limits which included a 200 nm zone around the rock (Figure 9.5). The Irish government duly protested against this action and, indeed, made a fisheries claim of its own in the same year which overlapped the British claim, not only in the rockall region, but also in the Celtic sea (Figure 9.5). In making this protest Ireland discounted not only Rockall, but all other British offshore islands not within three miles of the mainland, in delimiting the 'equitable equidistant line' between the two countries. (20)

The boundary line with Britain set out in Ireland's 1976 fisheries jurisdiction claim can also be taken as giving an indication of the Irish government's view as to the location of the seabed boundary line, although a small seabed designation make by Ireland off the northwest coast in 1977 demarcated a boundary line with Britain which departed somewhat from the fisheries designation of the previous year (Figures 9.1 and 9.5). This designation was apparently made in response to a decision by the British government to award hydrocarbon exploration licences in an area on the Irish side of where the Irish government considered the equidistant line between the two countries to lie. (21) The licensees in question (BP and BNOC) have not carried out any exploratory drilling in the area, and are unlikely to do so as long as the area's jurisdictional status remains in doubt.

In the Celtic Sea, the Irish government also designated a number of blocks which impinged upon Britain's 1976 fisheries designation in 1977

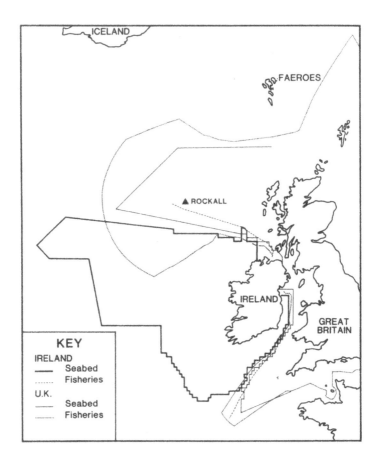

Figure 9.5. Offshore seabed and fisheries designations of Ireland and the United Kingdom

(Figures 9.1 and 9.5). This, again, was in response to a British seabed designation of the previous year which impinged upon the _Irish_ side of Ireland's 1976 fisheries designation. (22) This 1977 designation brought the areas designated by both sides for hydrocarbon exploration into direct contact (but not into overlap) for the first time.

IRISH-BRITISH BOUNDARY NEGOTIATIONS

It would appear that Ireland first approached Britain to submit the boundary question to arbitration in April 1976, and that Ireland's over-lapping seabed designation off the northwest coast in 1977 was designed to bring matters to a head. If this was so, it was certainly successful, for within three weeks the British government agreed to an independent arbitration procedure, although the official statement refers only to the continental shelf, and makes no reference to fisheries juris-diction. (23)

Thus far, negotiations on the arbitration procedure have been painfully slow: although it was agreed in 1980 that the arbitration board should be of an _ad hoc_ nature and would have five members, there has still been no agreement on its composition or terms of reference. The slowness of negotiations is perhaps surprising, given the growing interest (at least until the 1980 oil price collapse) in exploration in the Celtic and Irish Seas where, to date, some 70 wells have been drilled on the Irish side alone. Although neither side has made any further controversial designations since the decision to go to arbitration in 1977, it may be noted that in its ninth round of licensing awards (1985), the British government allocated two (previously-designated) Celtic Sea blocks which penetrate Ireland's fisheries limits. (24)

It would appear logical that any final boundary line between Ireland and the United Kingdom should be identical for both fisheries and seabed juris-diction. The overlapping Irish and British fisheries claims of 1976 have never been brought to a head because of the introduction soon after of the EEC Common Fisheries Policy which made boundaries between member states irrelevant. However, it would appear that Ireland will inevitably have to pursue its objection to Britain's declaration of a 200 nm fisheries limit around Rockall, in pursuit of its own seabed claims in the Rockall area; otherwise,

Ireland would be conceding the principle that uninhabitable rocks can generate exclusive economic zones of their own. Should Britain be forced to concede on this point, however, it would mean the loss to Britain - and thereby to all other EEC member states (including Ireland) - of a considerable area of rich fisheries territory to the west of Rockall. (25)

CONCLUSION

Ultimately, the boundary line between Ireland and the United Kingdom will depend on the status to be granted to Britain's offshore islands in the arbitration proceedings. In this context, Ireland will no doubt draw comfort from the 1977 arbitration decision on the respective British and French claims with respect to the English Channel/Western Approaches, whereby offshore islands were down-graded, and an 'equitable' solution was reached which divided the intervening area roughly half-and-half between the two sides. (26)

The division of the Atlantic region to the west of Ireland's and Britain's 200 nm limits between the four contending parties presents a much more complex set of considerations. While, as noted previously, there may be no immediate urgency as regards the delimitation of the respective national juris-dictions in this region, it is possible that, under the 1982 UNCLOS Convention, some of the area in question may properly belong to <u>none</u> of the disputants, in which case it would come under the aegis of the International Seabed Authority to be established under the Convention. It may, therefore, be in the interests of the disputants to arrive at an agreed division in advance of the possible intervention of the Boundary Commission provided for under UNCLOS, whose function it will be to decide the boundary lines between the areas of national and international jurisdiction. Whether such a pre-emptive division would hold its validity remains to be seen.

NOTES

1. The assistance of Piers Gardiner, Clive Symmons, Geraldine Skinner and Peter Innes in providing background information and commenting on earlier drafts of this paper is gratefully

acknowledged. All opinions, interpretations and errors are the sole responsibility of the author. James Keenan's cartographic assistance is also greatly appreciated.

2. P.R.R. Gardiner, 'Reasons and methods for fixing the outer limit of the legal continental shelf beyond 200 nautical miles', Iranian Review of International Relations, 11/12; (1978), 145-170.

3. Strictly speaking, a distinction may be made between offshore areas specifically designated (eg for hydrocarbon exploration or fisheries control) by a coastal state, and a more extensive area claimed as coming within that state's jurisdiction. This chapter is concerned mainly with offshore areas designated by the Irish government, but also considers the question of broader claims.

4. P.R.R. Gardiner and K.W. Robinson, 'The law of the sea', Technology Ireland, 9, 5, (1977), 7-12.

5. Ibid., 8.

6. C.R. Symmons, (1975) 'The Rockall dispute', Irish Geography, 8, (1975), 122-6.

7. P.R.R. Gardiner (1978), op. cit.

8. C.R. Symmons, 'The Rockall dispute deepens: an analysis of recent Danish and Icelandic actions', International and Comparative Law Quarterly, 35, (1986), 344-373.

9. The Law of the Sea, United Nations, (London, 1983).

10. These are the ultimate restrictions on offshore claims specified in Article 76; however, within these limits, there are a number of criteria which can be used to define offshore jurisdictions (eg Iceland's claim to territory in the Rockall-Faeroe Plateau).

11. C.R. Symmons (1986), op. cit.

12. As communicated personally to the author.

13. It is important to distinguish between geographical and legal definitions of the continental shelf; the latter, as specified in the

1982 UNCLOS Convention includes the slope and rise (see above) as well as the geological shelf.

14. See _Irish Times_, May 8, 1985, 5.

15. See _Irish Times_, May 25, 1985, 20.

16. _Irish Times_, September 10, 1974, 1.

17. _Ibid._

18. C.R.Symmons (1975), _op. cit._

19. See _Irish Times_, May 8, 1985, 5.

20. C.R. Symmons, 'The outstanding maritime boundary problems between Ireland and the UK' Paper to the 19th Conference of the Law of the Sea Institute, (Cardiff 1985), see also _Irish Parliamentary Debates_, February 16, 1977: Col. 1572.

21. See _Irish Times_, February 3, 1977, 4 and February 4, 1977, 1.

22. C.R.Symmons (1985), _op. cit._

23. See _Irish times_, February 23, 1977, 1. Also C.R.Symmons (1985), _op. cit._

24. C.R. Symmons (1985), _op. cit._

25. _Ibid._ See also _Irish Parliamentary Debates_, February 16, 1977: Col. 1574. It may be noted that, according to the UNCLOS Convention, there is no provision for the extension of fisheries rights beyond the 200 nm Exclusive Economic Zone, unlike in the case of seabed resources.

26. C.R. Symmons (1985), _op. cit._

CHAPTER 10

MARITIME BOUNDARY PROBLEMS IN THE BARENTS SEA

R.R. Churchill

INTRODUCTION

The aim of this chapter is to examine the continuing dispute between Norway and the USSR over the
boundary between their respective continental
shelves and 200-mile economic zones in the Barents
Sea. The origin of the dispute will be described,
and an account given of the temporary solution to
the problem of the absence of an economic zone
boundary. An attempt will then be made to evaluate
the position taken by each party in the dispute in
the light of the applicable international law.
First, however, it is necessary to say something
about the resources of the Barents Sea, the management and allocation of which will be directly
affected by any boundary which is eventually agreed,
and to explain the strategic significance of the
area, which largely accounts for the lack of
progress so far in solving the boundary dispute.

THE BARENTS SEA: RESOURCES AND STRATEGIC
SIGNIFICANCE

The Barents Sea is bounded to the south by the
mainland coasts of Norway and the USSR, to the east
by the large Soviet archipelago of Novaya Zemlya,
and to the north by the archipelagos of Franz Josef
Land and Svalbard, belonging to the USSR and Norway
respectively. In the west the Barents meets the
Greenland Sea, the conventional dividing line being
taken as the line running from the South Cape (on
Svalbard), through Bear Island (Norwegian) to the
North Cape on the Norwegian mainland. Thus defined,
the Barents Sea is about 1.4 million square kilometres (about 542,000 square miles) in area. (1) It

is a comparatively shallow sea, only about half being deeper than 200 metres. The deepest point is a little over 500 metres. The average depth is 229 metres. (2)

The Barents is very rich in living resources. The most important commercial species are cod, haddock, saithe, redfish, capelin and shrimp. In recent years the total catch has been between about 1 and 1.5 million tonnes per annum, some two-thirds of which comprises capelin (used mainly for reduction to meal and oil). Over 95 per cent of the catch is taken by Norwegian or Soviet vessels. There is also a large seal population in the Barents, the harvesting of which has substantially decreased in recent years following the decline in the demand for sealskin furs.

The bed of the Barents Sea is believed to contain considerable quantities of oil and gas. In spite of a certain amount of exploratory drilling, no finds have yet been made (or at least have not been publicly announced as having been made).

In addition to the wealth of its natural resources, the Barents has a very great strategic significance. The Kola peninsula, on the southern shores of the Barents, is the base of the USSR's Northern Fleet, the largest and most important of its four navies. Furthermore, the Northern Fleet is the only Soviet navy with relatively unconstricted access to the main oceans of the world. Not only does the Barents contain the base of the USSR's most important navy, it has now become one of the main operating areas for Soviet nuclear submarines: from the Barents nuclear warheads from submarines can reach most of North America and Western Europe. This has increased the strategic significance of the Barents not only for the USSR, but also for the USA and its NATO allies. Moreover, because the Kola base, being concentrated in a small area only 100km from a NATO country (Norway), is vulnerable to a surprise attack, and because of naval movements in and out of the Kola peninsula, the USSR is extremely sensitive to any non-Soviet activities in neighbouring land and sea areas. (3)

ORIGINS OF THE DELIMITATION DISPUTE

Under the 1958 Continental Shelf Convention (to which both Norway and the USSR are parties) the continental shelf is defined as the seabed out to 200 metres or beyond if the depth of the superjacent waters admits of the exploitation of the resources of the seabed. Under the 1982 UN Convention on the Law of the Sea the continental shelf is defined as the seabed out to 200 miles or the edge of the continental margin, whichever is the further. Although the 1982 Convention is not yet in force, the International Court of Justice has suggested that the definition of the continental shelf by the Convention has already passed into customary law. (4) Thus under the 1958 Convention at least part of the bed of the Barents Sea is continental shelf - that part lying in less than 200 metres of water, together with however much beyond that which may be exploitable. Under the UN Convention and customary law there is no doubt that the whole of the bed of the Barents Sea is, legally speaking, continental shelf. This means that since the emergence of the continental shelf as a concept in international law in the 1950s, there has been the need to establish a boundary between the continental shelves of Norway and the USSR - initially perhaps for only part of the Barents, now undoubtedly for the whole Sea.

Norway has taken the view that the boundary should be the median line (ie a line equidistant from the nearest part of each state's mainland or insular territory). The USSR has argued (at least in the earlier stages of the dispute) that the boundary should be a sector line, ie the line of longitude running from the terminus of the Norway-Soviet land frontier to the North Pole, modified so as to avoid passing through the Svalbard archipelago (Figure 10.1). The seabed lying between the median line and the sector line is an enormous area of approximately 45,000 square nautical miles - an area greater than the Norwegian sector in the North Sea. Since 1974 several rounds of negotiations over the course of the boundary have taken place without apparently a great deal of narrowing of each side's position, although the Norwegian Government has made it clear that it would be prepared to modify its position on the median line in return for some concessions on the USSR's sector claim. So far the USSR does not appear to have indicated any willingness to make such concessions. The justification for each party's position will be examined in further detail below.

Since 1977 negotiations over a continental shelf boundary have become further complicated by the establishment of 200-mile economic zones by both Norway and the USSR. (5) This development meant that future negotiations would be concerned not just with

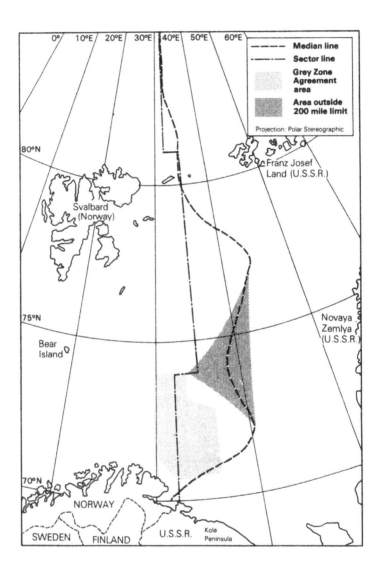

Figure 10.1. Maritime boundaries in the Barents Sea

a continental shelf boundary but also with an economic zone boundary. In the late 1970s there was no immediate urgency for a continental shelf boundary, as neither Norway nor the USSR desired to begin early exploration for hydrocarbons in the disputed area. The same was not true for an economic zone boundary, however, as the Southern Barents has been long and heavily fished by both Norwegian and Soviet fishermen. Neither Norway nor the USSR wished speedily to establish an economic zone boundary (nor would such a boundary have been readily agreed), as this would largely have pre-empted the negotiations over the continental shelf boundary. Both sides were agreed, however, on the need to come to some temporary practical arrangement speedily for the exercise of each State's fisheries jurisdiction in the waters lying over the disputed area of continental shelf ie the area between the median line in the east and the sector line in the west and within 200 miles of the mainland. This shared point of view led to the signature of the so-called 'Grey Zone Agreement' in January 1978. (6)

TEMPORARY SOLUTION TO THE ECONOMIC ZONE BOUNDARY

The agreement applies to an area which not only covers a large part of the disputed continental shelf area in the Southern Barents (the grey zone proper), but also to areas which are indisputably Norwegian or Soviet, ie areas to the west of the sector line and east of the median line respectively (see the map at the end). Within this area total allowable catches for the various fish stocks are to be decided by the Norwegian-Soviet Fishery Commission established by an earlier fishery agreement of 1975. Total allowable catches are then to be allocated roughly equally between the two parties, with some quotas being allocated to third States after mutual consultation. The Agreement also sets out various other regulatory measures (eg as to fishing gear and minimum fish sizes) which are to be observed by all vessels fishing in the area. Finally, the Agreement provides that each party may exercise jurisdiction only in respect of its own fishing vesels and not in respect of vessels of the other party: jurisdiction over the fishing vessels of third States is to be exercised by whichever party has licensed such vessels.

The Agreement is a temporary one, and was originally concluded for one year only. It has subsequently been extended for annual periods in every year following its signature, and this is likely to continue until agreement is reached on a continental shelf/economic zone boundary. In general the Agreement has worked well, and it is some measure of the Agreement's perceived usefulness and success that its annual renewals have continued uninterrupted notwithstanding the fact that the non-socialist parties in Norway who were originally opposed to it, replaced the Labour Party as the Government and the fact that a prominent member of the Norwegian delegation that negotiated the Agreement, Arne Treholt, was convicted of being a Soviet spy in 1985.

Having examined the origins of the dispute over the continental shelf and economic zone boundaries (and the temporary solution to the absence of an economic zone boundary), we must now turn to consider the arguments of the parties as to where the boundary should run in the light of the applicable international law.

THE ARGUMENTS OF THE PARTIES IN THE LIGHT OF THE APPLICABLE INTERNATIONAL LAW

At the outset it is desirable to recognise that the disputed boundary has three distinct segments (though whether the parties so view it is not known). The first segment runs from the termination of the existing maritime boundary (largely a territorial sea boundary) (7) at the mouth of Varangerfjord to a point 200 miles from the mainland of either Norway or the USSR (or possibly a point 200 miles from both). In this segment there is a need for both a continental shelf and economic zone boundary between the adjacent mainland coasts of Norway and the USSR. The second segment comprises the area in the middle of the Barents Sea which is more than 200 miles from any land. Here only a continental shelf boundary is required. In this segment the relevant coasts of Norway (in part Svalbard, in part the mainland and the USSR (Novaya Zemlya) are essentially opposite one another. In the final segment, in the Northern Barents, there is need once again for both a continental shelf and economic zone boundary. Here the relevant coasts – Novaya Zemlya and Franz Josef Land on the Soviet side, Svalbard on the Norwegian side – are again opposite. Each of the three segments must be considered separately, both because the applicable

law differs and because different factors relevant
to delimitation apply to each.

We will begin with the southern segment, which
in practice is the area where the need for a
boundary is greatest. Before 1977 negotiations in
this area were concerned solely with a continental
shelf boundary. For this the applicable law was the
Continental Shelf Convention (since both states are
parties to it). Article 6 of the Convention provides
that in the absence of agreement, the boundary line
is to be the median line unless another line is
justified by special circumstances. Since 1977 it
appears (though like much else in the negotiations,
little has been said publicly) that the negotiations
have become concerned with seeking to establish a
single boundary for both the economic zone and
continental shelf.(8) If this is so, then according
to the International Court of Justice in the Gulf of
Maine case (9) the applicable law is no longer
Article 6 of the Continental Shelf Convention but
customary international law. According to the Court,
customary law lays down the basic principle that
'delimitation is to be effected by the application
of equitable criteria and by the use of practical
methods capable of ensuring, with regard to the
geographic configuration of the area and other
relevant circumstances, an equitable result', (10)
but does not prescribe any particular 'criteria' or
'practical methods' for effecting a delimitation.

Norway takes the view that the most appropriate
practical method for effecting a delimitation is the
median line (or, in terms of the Continental Shelf
Convention, that there are no special circumstances
that justify a departure from the median line). The
USSR takes the view that there are factors which
indicate that a median line is not the appropriate
method of delimitation (or, in terms of the
Continental Shelf Convention, that there are special
circumstances). The factors that the USSR apparently
has in mind are its greater size, the greater
population of the Kola peninsula as compared with
Northern Norway, and its security interests. (11) As
regards the first of these, this is not a factor
that an international court would be likely to
consider relevant: in the Continental Shelf Libya/
Malta case, for example, the International Court of
Justice rejected Libya's argument that a State with
a greater land mass should have a greater share of
continental shelf. (12) However, if by its greater
size what the USSR is really getting at is the fact
that its coastline in the Southern Barents is

considerably longer than that of Norway, each coast
being measured according to its general direction
and not in all its sinuosities (as international
courts prescribe), then it is on much firmer legal
ground. International courts have stressed several
times that there should be a reasonable (though not
exact) degree of proportionality between the maritime areas delimited and the length of the parties'
respective coastlines. (13) Thus, it seems that the
USSR would be justified in raising the greater
length of its coastline as a relevant factor to be
taken into account in the delimitation. The other
factors alleged by the USSR are its greater
population, and security. The trend of the
International Court's decisions is to exclude as
irrelevant to delimitation all factors except the
geographical, particularly where, as is likely to
the case here, the boundary line relates to both the
continental shelf and economic zone. On the other
hand, in both the Guinea/Guinea Bissau arbitration (14) and the Continental Shelf Libya/Malta
case,(15) security factors were not dismissed out of
hand as irrelevant, though in neither case was it
made very clear in what circumstances they might be
relevant: what the court in each case seems to have
had in mind is that a state should have control of
the maritime territory immediately off its coast. In
that sense, security factors would seem to reinforce
the non-encroachment principle (discussed below).

There is a further factor which has not
apparently been raised by the USSR but which might
be relevant. The Norwegian coast projects a little
further seawards than the Soviet coast along what is
basically a straight coast between the White Sea and
the North Cape. The effect of this is to push the
median line a little further to the east, possibly
to such an extent that it might be regarded as
encroaching on maritime areas that more naturally
belong to the USSR. The principle of non-
encroachment is one that has been widely recognised
by international courts. (16)

It was pointed out earlier that the USSR has
apparently argued that the continental shelf
boundary should be a sector line. The actual line of
longitude to which the USSR refers was that used in
a Soviet decree of 1926 to define the western limit
of the USSR's claim to land (not maritime) territory
in the Arctic. It is not clear whether the USSR
regards the sector line as the line which results
from the existence of the special factors discussed
above or whether the sector line itself is regarded

as a relevant factor. The former seems unlikely, as it would be highly artificial to consider that the relevant factors should give rise to such a precise and, from the Soviet point of view, well-established line. The latter possibility also seems unconvincing, for a sector line is not of the character of the factors which international courts have hitherto recognised as relevant. It is also possible that the USSR considers that use of the sector line has a justification quite separate from the ordinary law of maritime boundary delimitation, namely that a sector line is the appropriate way of drawing a maritime boundary in polar areas. If this is the Soviet contention, the USSR would have to prove that the use of sector lines as a method for drawing maritime boundary lines in polar areas is a rule of customary international law. This would seem impossible to prove, because there is both insufficient practice and inadequate opinio iuris (State practice and opinio iuris being the two constituent elements of customary international law). It is true that sector lines have been used as boundaries in the Antarctic, but this use is limited to land territory, and seems to be motivated by reasons of convenience, not because the states concerned felt themselves to be under any legal obligation to use such lines. (17) In the Arctic, Norway and the USA have protested against the use of sector lines, and USSR is the only state which has used such lines. (18) In any case an international court would probably regard the sector line as being an inequitable solution in the Barents because it would encroach on maritime areas which more properly belong to Norway.

In relation to the first segment of the maritime boundary in the Barents, the most important factors would seem to be that Norway and the USSR are adjacent states along what is essentially a straight coastline and that the Soviet coastline in the Southern Barents is considerably longer than the Norwegian coastline. In this situation neither the median line nor the sector line appear to be the appropriate method or constitute an equitable solution. What might be regarded as an equitable solution would be to draw the boundary line perpendicular to the general direction of the coast. This was commended by the International Court in the Gulf of Maine case as a suitable method of delimitation of coasts which are more or less straight. (19) Furthermore, a perpendicular line would respect the principles of non-encroachment and proportionality.

Having looked at the first segment, we must now turn to look at the second, in the middle of the Barents. Here, it will be recalled, only a continental shelf boundary is required. Thus the applicable law will be Article 6 of the Continental Shelf Convention. In other words, the boundary will be the median line unless another line is justified by special circumstances. Of the factors raised by the USSR and discussed above, the sector line is clearly not a special circumstance. Nor is the principle of non-encroachment relevant where, as here, the respective coasts are opposite one another rather than adjacent. While it is difficult to calculate the length of the relevant Norwegian coastline (part of Svalbard, and possibly part of the mainland), it does not seem excessively disproportionate when compared to the relevant Soviet coastline (part of Novaya Zemlya). What could possibly amount to a special circumstance, and therefore require some adjustment of the median line, is the fact that two of the islands in the Svalbard archipelago, Hope Island and Bear Island (and the latter may not be relevant), lie a long way from the main part of the archipelago.

In the third and final segment there is, as with the first segment, the need for both a continental shelf and economic zone boundary. Thus, the applicable law will be customary law, and not the Continental Shelf Convention. Unlike in the first segment, the coasts of the parties here are opposite one another, not adjacent. This means that the principle of non-encroachment would not seem to be relevant. Furthermore, again unlike in the first segment, there does not seem to be an excessive disproportion between the lengths of the parties' coastlines. In general international courts have taken the view that equidistance is an appropriate method for delimitation (at least as a starting point) in the case of opposite coasts. (20) Thus it would seem that in the present case equidistance would in principle be the appropriate method of delimitation, though some adjustment might be thought desirable to discount (wholly or in part) some of the outlying islands of the Svalbard and Franz Josef Land archipelagos.

TOWARDS A SOLUTION?

The means by which a solution to the boundary

dispute in the Barents may be reached are either continued negotiations between the parties or recourse to an international judicial or arbitral tribunal. The latter is most unlikely, given the USSR's longstanding opposition to the use of such tribunals for solving disputes in which it is involved. Thus, it seems that continued bilateral negotiations will be the medium used to try to reach a solution. It should be stressed that while some suggestions have been made above as to the possible course of the boundary indicated by international law, it is open to the parties, whether the applicable law is the Geneva Convention or custom, to agree on any boundary line they wish. The discussion of the law in the preceeding section, particularly the case law of international courts, is not, however, without interest or relevance. It is likely that various aspects of this law are quoted by the parties in the negotiations in order to support a particular line of argument.

The speed at which negotiations over a maritime boundary in the Barents are pursued and the likelihood of their reaching a successful conclusion depend to a considerable extent on how strongly the parties feel the need to explore new areas for seabed hydrocarbons. (As long as the 'grey zone' agreement remains in force, fisheries interests will not press either party to seek an early solution to the boundary problem). In 1986 both Norway and the USSR appear to be well endowed with hydrocarbon resources elsewhere, nor is the Barents yet apparently of any great commercial interest to the oil companies. However, it may be significant that since 1983 the USSR has conducted exploratory drilling just to the east of the disputed area. It is also noteworthy that at their last round of negotiations, in December 1984, the parties apparently agreed that they would not conduct exploratory activities within the disputed area so long as a boundary is not agreed. (21) Experience elsewhere in the world (particularly in the North Sea) suggests that it will be easier for Norway and the USSR to agree on a continental shelf boundary if the hydrocarbon resources of the disputed area are not identified beforehand. While the respective needs of the parties for hydrocarbons is a major factor in determining the course of negotiations, it is not the only factor. For the USSR security considerations are also very important. Its security interests would, in the absence of any pressing need for resources, lead the USSR to delay negotiations.

So long as there is no agreed boundary, the Norwegians will not drill in the disputed area or in all probability go very close to it. Wherever a boundary is located, it will almost certainly lead to Norwegian offshore activity pushing eastwards, something which is inimicable to Soviet security interests: the further east Norwegian offshore activity takes place, the greater the possibility of Norwegian drilling rigs obstructing or monitoring Soviet naval vessels on their way into or out of the ports of the Kola peninsula. A final factor in determining the speed of negotiations is the fact that the USSR may perceive that Norway appears to be keener on a boundary agreement than it is: it can therefore afford to wait, hoping that this will lead to concessions by Norway.

If the negotiations achieve a positive result, it will probably be in the form of agreement on a single boundary line. A theoretical alternative is to abandon the search for a boundary and to provide instead for a regime of joint user for the disputed area (as, for example, Sudan and Saudi Arabia have done in the Red Sea). It seems unlikely that such a solution would appeal to either of the parties. If the solution is, as is most likely, a single boundary line, it may possibly be more readily achieved if the boundary agreement were to provide for the non-use of offshore installations for espionage purposes (although this may raise difficult questions of verification and inspection) and for non-obstruction by installations of either party's vessels. For the sake of fisheries interests and the environment generally, it is to be hoped that a boundary agreement will continue the present arrangements for the management of joint fish stocks and contain meaningful provisions on pollution control.

CONCLUSIONS

From a strictly geographical point of view, the Barents Sea, especially in the South, poses no particularly difficult problems for maritime boundary delimitation: indeed, compared with many areas, it would seem a fairly straightforward case. The fact that after eleven years of intermittent negotiations Norway and the USSR appear to be nowhere near reaching agreement on a maritime boundary is due very largely to the strategic and security factors described at the beginning of this paper. Of the 200 or more major unresolved maritime boundaries around the world, the Barents is probably unique on the extent to which strategic and security considerations are a factor in determining the boundary. From this point of view the Barents is therefore of rather limited interest to students of maritime boundary delimitation.

Of much greater, and very real interest is the 'grey zone' agreement - the temporary (though now eight-year old) solution to the problem of the absence of a boundary in respect of jurisdiction over living resources. This agreement demonstrates how living resources can be effectively managed and allocated in an area where a maritime boundary is disputed. It may therefore have lessons to offer other parts of the world where boundaries concerning jurisdiction over living resources are in dispute.

NOTES

1. K. Traavik and W. Ostreng, 'Security and Ocean law: Norway and the Soviet Union in the Barents Sea' Ocean Development and International Law, 4: (1977) 343-367.

2. Ibid.

3. For further discussion of the strategic significance of the Barents, see K. Traavik and W. Ostreng, (1977), op.cit, 351-53 and C. Archer and D. Scrivener 'Frozen Frontiers and Resource Wrangles: Conflict and Co-Operation in Northern Waters' International Affairs 59: (1983) 59-76, 69-70.

4. Case concerning the Continental Shelf (Libya/Malta) (1985) ICJ Rep. 13 at 33. In its domestic legislation the USSR uses the Continental Shelf Convention definition (see Decree of February 6, 1968: text in UN Legislative Series B/15, p.441). Norwegian law originally defined the continental shelf as being the seabed out to the depth of exploitability (see Royal Decree of May 31, 1963: text in ibid., 393), but the definition has now been changed to that of the UN Convention (see the Petroleum Law of March 22, 1985).

5. The Norwegian economic zone was established by Law No. 91 of December 17, 1976, text in UN Legislative Series B/19, 241. The Soviet Union originally established a 200-mile fishing zone (see

Decree of December 10, 1976, text in ibid,. 253), but replaced this in 1984 with a 200-mile economic zone (see Decree of February 28, 1984, text in Law of the Sea Bulletin No. 4 (United Nations 1984), 32).

6. The official title of the Agreement is: Agreement on an Interim Practical Arrangement for Fishing in an Adjoining Area in the Barents Sea. An English translation of this Agreement does not appear to have been published: the Norwegian text can be found in Overenskomster med fremmede Stater (1978) No. 436.

7. Established by the Agreement concerning the Sea Frontier between Norway and the USSR in the Varangerfjord, 1957. Text in UN Treaty Series, Vol. 312, 289.

8. This is certainly the implication given by the then Norwegian Prime Minister in a speech made in February 1978: see UD - informasjon, 7, (1978).

9. ICJ Report (1984) No. 246.

10. Ibid, 299-300.

11. As mentioned above, few details of the negotiations have been made public. The factors mentioned here have been indicated by the Norwegian Government as being the ones the USSR has in mind: see UD - informasjon 30 (1977).

12. Case concerning the Continental Shelf (Libya/ Malta) ICJ Report 13 (1985) at 40-41. Note, too, that in the Guinea/Guinea Bissau Maritime Boundary Arbitration (text reproduced in Revue Generale de Droit International Public 89: (1985), 484-537 the Court said that the respective size of the parties' land territory was irrelevant (paras. 118-119).

13. See North Sea Continental Shelf cases ICJ Report 3 (1969) at 52; Continental Shelf (Tunisia/ Libya) case ICJ report 18 (1982) at 75; Gulf of Maine case ICJ Report 246 (1984) at 322-323, 334-337; Guinea/Guinea Bissau Arbitration, paras. 118-120; Continental Shelf (Libya/Malta) case ICJ report 13 (1985) at 43-46, 49-50.

14. Guinea/Guinea Bissau Arbitration, (1985), para. 124.

15. ICJ Report (1985) 13 at 42.

16. See North Sea Continental Shelf cases ICJ
Report 3 (1969) at 31-32; Continental Shelf
(Tunisia/Libya) case ICJ report 18 (1982) at 61-62;
Gulf of Maine case ICJ Report 246 (1984) at 298-9,
313, 328; Guinea/Guinea Bissau Arbitration, paras.
103-107.

17. D.J. Harris Cases and Materials on International Law, (London, 3rd Edition, 1983).

18. Ibid, 183. Canadian practice concerning sector
lines in the Arctic is ambivalent. See D. Pharand
The Law of the Sea of the Arctic with Special
Reference to Canada (University of Ottowa, Ottawa,
1973), 134-141, 1979, D. Pharand 'The Implications
of Changes in the Law of the Sea for the 'North
American' Arctic Ocean' in J.K. Gamble (ed) Law of
the Sea: Neglected issues (Law of the Sea Institute,
Hawaii, 1979), 184.

19. ICJ report 246 (1984) at 320. A similar view
appears to have been taken by the Court in the
Tunisia/Libya case.

20. See the North Sea Continental Shelf cases ICJ
Report 3 (1969) at 36-37; Anglo-French Continental
Shelf Arbitration, Cmnd. 7438, para. 239;
Continental Shelf (Tunisia/Libya) case ICJ Report 18
(1982) at 88; Gulf of Maine case ICJ report 246
(1984) at 334; Continental Shelf (Libya/Malta) case
ICJ Report 13 (1985) at 47.

21. The Times, December 4, 1984. Whether international law requires a State to abstain from
drilling in a disputed area of continental shelf is
not a question to which a straightforward answer can
be given.

Acknowledgement

The author is grateful to the Cartographic Unit, of
the Department of Maritime Studies at UWIST for
drawing Figure 10.1.

CHAPTER 11

THE UNITED STATES EXCLUSIVE ECONOMIC ZONE: MINERAL RESOURCES

Fillmore C.F. Earney

INTRODUCTION

On 10 March 1983, United States President Ronald Reagan issued a policy and action statement – Proclamation 5030 – which established a national oceanic Exclusive Economic Zone (EEZ). The EEZ extends 370 km (200 nautical miles, or nm) seaward from the baseline used to measure the 5.6 km (3 nm) territorial sea (Figure 11.1). Within the EEZ, the United States (US) claims sovereignty over all resources of the water column (except certain migratory species), the seabed, and the subsoil.(1)

President Reagan's proclamation added 1.255 milliard hectares (3.1 billion acres) to the 0.931 milliard hectares (2.3 billion acres) of land area (including states and territories) already administered by the US Government (Figure 11.2). This action has been compared to the 1803 US purchase, from Napoleonic France, of Louisiana, a vast territory lying between the Mississippi River and the Rocky Mountains that included a bounty of untapped resources. Like the Louisiana Purchase, establishment of the EEZ presents the US Government with significant challenges; it must: 1) assess the resources it has acquired; 2) develop programmes to make these resources available to industry; and 3) negotiate numerous new international boundaries. These challenges are the main focus of this paper, especially as they relate to the EEZ's mineral resources.

RATIONALE AND PRECEDENT

The US is highly dependent on foreign sources of minerals (Table 11.1). Reducing this dependence

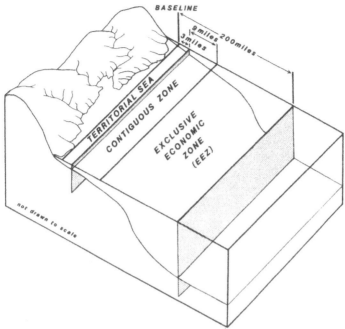

Figure 11.1. Schematic diagram of the baseline, territorial sea, contiguous zone, and EEZ. Source: R.W. Rowland, M.R. Goud, and B.A. McGregor, The US Exclusive Economic Zone - A Summary of Its Geology, Exploration, and Resource Potential. USGS Circular 912, (Alexandria, Va. 1983), p.3

has been a major concern of the Reagan Administration. On 5 April 1982, President Reagan issued a statement outlining a national minerals policy where he said: 'It is the policy of this Administration to decrease America's minerals vulnerability by taking positive action that will promote our national security, help ensure a healthy and vigorous economy, create American jobs, and protect America's natural resources and environment.' (2). Establishment of the EEZ was, in his view, a 'positive step' toward reducing the US' minerals vulnerability. Many observers did not share his perception.

The US has been castigated for unilaterally declaring an EEZ, being branded an international renegade set on destroying the viability of the 1982 United Nations Conference on the Law of the Sea (UNCLOS III). After a decade of negotiation, on 10

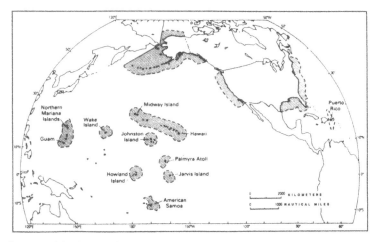

Figure 11.2. United States EEZ. Source: D.L. Peck, p.57, (see note 29)

December 1982, 117 world states signed the Convention, with several others signing by 10 December 1984, the date of closure. As of early 1986, however, little more than one third of the states needed to ratify to bring the Convention into force had done so. The Convention will not take effect until twelve months after the sixtieth ratification. The US, along with several other industrial states, did not sign the Convention. President Reagan has stated many times that, because of the Convention's failure to meet the deep seabed interests of the US and its mining industries, the US (under his administration) will not become a Party.

In reality, the United States EEZ Proclamation is only an extension of its Magnuson Fisheries Conservation and Management Act of 1976 (FCMA) that took effect 1 March 1977, establishing a 370-km fishery zone. Under this act, foreign fleets operating within 370 km of the coast must obtain licences, take only a specified tonnage of fish, and submit to inspection by the US Coast Guard and the US National Marine Fisheries Service. An earlier precedent occurred in 1945 when President Harry S. Truman issued a proclamation laying claim to seabed and subsoil resources out to the edge of the continental shelf.(3) Other states viewed this unilateral action as giving them a legitimate right to protect their vital offshore resources. In 1947,

Chile and Peru declared a patrimonial control over fisheries out to 370 km off their shores, followed in 1951 by Honduras and in 1952 by Ecuador. During the late 1950s and 1960s, other states made claims to varying distances in their offshore waters, as in the celebrated Icelandic fishery disputes.

Table 11.1 United States net-import reliance for selected minerals potentially available from the oceans, 1985

Mineral	Per cent
Manganese	100
Cobalt	95
Platinum-group metals	92
Chromium	73
Zinc	69
Nickel	68
Silver	64
Vanadium	54
Gold	31
Copper	27
Iron ore	22
Lead	16
Sulfur	5

Source: US Bureau of Mines, Mineral Commodity Summaries 1986, (Washington DC, 1986), pp.2, 88, 172; data for lead and vanadium are for 1984.

When the first session of UNCLOS III convened in 1973, the 370-km EEZ had become a generally accepted national prerogative; during the session 110 states favoured and only eight opposed the concept.(4) By 1976, 17 states had unilaterally declared EEZs, followed in 1977 by another 16 states (including the US with its fishery zone). As of early 1986, 59 states had established and put into force EEZs (Table 11.2). Several other states have declared EEZs but have not enacted or put into force implementing legislation. In addition, 12 other states claim 370-km territorial seas; two states (Argentina and Chile) claim the continental shelf and slope out to 648 km (350 nm) and a third (Ecuador) claims the seabed between its mainland and its Galapagos Islands, lying 1000 km (540 nm) offshore. Still another 22 states have declared

370-km fishery zones. In total, nearly 100 states make some form of 370-km offshore claim. Thus, the US cannot be accused of circumventing customary international law.

Table 11.2 Coastal states that have unilaterally declared and put into force an exclusive economic zone (1)

Antigua and Barbuda	Guatemala	Pakistan
Bangladesh	Guinea	Philippines
Barbados	Guinea-Bissau	Portugal
Burma	Haiti	Samoa
Cape Verde	Honduras	Sao Tome and Principe
Colombia	Iceland	Senegal(2)
Comoros	India	Seychelles
Cook Islands	Indonesia	Solomon Islands
Costa Rica	Ivory Coast	Spain
Cuba	Kenya	Sri Lanka
Democratic Kampuchea	Madagascar(2)	Suriname
Democratic People's Republic of Korea	Mauritania	Togo
	Mauritius	Tonga
Democratic Yemen	Mexico	Union of Soviet
Djibouti(2)	Morocco	Socialist Republic(2)
Dominica	Mozambique	United States of America
Domincan Republic	New Zealand	Vanuatu
Equatorial Guinea(2)	Nigeria	Venezuela
Fiji	Niue	Viet Nam
France	Norway	
Grenada	Oman	

Sources: (1) Office of the Special Representative of the Secretary-General for the Law of the Sea, _Law of the Sea Bulletin_ No.2 (New York), March 1985, ii-iv; (2) A. Demarffy, Letter (14 Jan. 1986) Senior Officer, UN (New York).

CONSISTENCY

If the US cannot be cast in ·the role of an international misfit and if, as the Reagan Administration claims, it seeks to adhere to the prescriptions and spirit of the LOS Convention (with the exception of its mandates for the administration of the deep seabed), how consistent are the assertions of Proclamation 5030 with Convention articles that deal with the EEZ?

For the most part, assertions of the Proclamation and the articles of the Convention are consistent. The two documents outline nearly identical rights and duties. The Proclamation does diverge slightly from language contained in the Convention's Article 56 (the second of 21 articles dealing with the EEZ) titled 'Rights, jurisdiction and duties of the coastal state in the exclusive economic zone' which reads:

1. In the exclusive economic zone, the coastal State has:
(a) sovereign rights for the purpose of exploring and exploiting, conserving and managing the natural resources, whether living or non-living, of the waters superjacent to the sea-bed and its subsoil, and with regard to other activities for the economic exploitation and exploration of the zone, such as the production of energy from the water, currents and winds;
(b) jurisdiction as provided for in the relevant provisions of this Convention with regard to:
  i) the establishment and use of artificial islands, installations and structures;
  ii) marine scientific research;
  iii) the protection and preservation of the marine environment;
(c) other rights and duties provided for in this Convention.
2. In exercising its rights and performing its duties under this Convention in the exclusive economic zone, the coastal State shall have due regard to the rights and duties of other States and shall act in a manner compatible with the provisions of this Convention. (5)

Whereas the Convention gives coastal states the right to control marine research activities, the US Proclamation does not. Indeed, the US seeks the fullest possible access for scientific researchers to all ocean waters (whether coastal or deep sea), but it also affirms the right of other states to monitor marine research activities in their EEZs.

More significantly, the Convention (Article 76) drops the use of the indeterminate 'technological exploitability' criterion for coastal state offshore jurisdiction (as used by the 1958 Geneva Convention on the Continental Shelf), still adhered to by the US. The UNCLOS III Convention sets the EEZ's outer limit at 370km (200 nm) from the territorial sea baseline or at the outermost edge of the continental margin, whichever is wider, with an absolute limit of 648km (350 nm). (6)

Other textual divergences occur in relation to exclusive management control of anadromous species of fish throughout their migratory range, even beyond the 370km limit. The US contends that international agreements should control anadromous fish within as well as outside the EEZs, whereas the Convention disallows this authority within the EEZ (Article 66). (7) Finally, under the Convention, coastal states are to provide geographically disadvantaged states access to an equitable part of the catch surplus (Article 70). (8) Under the United States FCMA, the allocation is based on cooperative reciprocity and trade policies. In addition, non-compliance to fishing regulations under the FCMA provides for imprisonment; under the Convention, incarceration is specifically forbidden (Article 62). (9)

Although no major differences occur between the Convention and the United States oceanic legislation and its Proclamation 5030, some international law specialists feel that the US, not to be contradictory in its EEZ management regime and its claim to conformance with the Convention, must amend some legislation. If the US fails to act, other states may issue legislation that significantly deviates from the Convention mandates for the EEZ, thus weakening the likelihood of a stable and comprehensive oceanic administrative regime. (10) A 1984 report to The President and The Congress by the US National Advisory Committee on Oceans and Atmosphere presents this view. The Committee recommended that no comprehensive EEZ legislation be implemented, but suggested that specific non-conforming legislation should be carefully reviewed and updated, using as a guide a 1978 government study 'Federal Legal Responsibilities in a Potential 200 Mile Economic Resource Zone'. (11)

BOUNDARIES

The EEZ Proclamation 5030 reconfirms boundaries established in the 1977 FCMA, which created more than 20 new maritime boundaries with adjacent and opposite states. Opposite boundaries become an issue when the distance between states is less than 740km (400nm). New boundaries or extensions of old boundaries requiring negotiation include: four with Canada, possibly three with Kiribati, three with Mexico, and one each with Anguilla (UK), the Bahamas, Cuba, British Virgin Islands, Dominican Republic, the Netherlands Antilles, the Federated States of Micronesia, Marshall Islands, Tonga, Western Samoa, Cook Islands, New Zealand (Tokelau), and Japan. With future development of outer portions of the EEZ for minerals, delimiting many of these boundaries will become increasingly important. (12)

CANADA

Four maritime boundaries shared by Canada and the US were under negotiation prior to the establishment of the FCMA or EEZ. One, in the Gulf of Maine, was settled on 12 October 1984, through binding adjudication by the International Court of Justice. (13) Machias Seal Island and North Rock in the northern Gulf remain in dispute. (14) At issue in the Gulf of Maine case were fisheries and potential petroleum resources. Negotiations on the other three boundaries - the Beaufort Sea between Alaska and Yukon Territory, the Dixon Entrance between islands of Alaska and British Columbia, and the Strait of Juan de Fuca between Washington and British Columbia - were to begin once the Gulf of Maine case was settled.

In the Beaufort Sea boundary dispute, Canada insists on a northward extension of the 141°W. meridian, whereas the US argues for an equidistant line struck at a right angle to the coastal configuration. The difference in the position of these two lines creates a significant wedge of territory with good potential for petroleum resources (Figure 11.3). The case of the Dixon Entrance centres upon the status of the Strait's waters. Canada claims all of the waters, whereas the US opts for an equidistant maritime division. The differing views stem from varying interpretations of the Alaska Boundary Tribunal Award of 1903 and earlier treaties. The Juan de Fuca dispute involves only minor discrepancies in charting methodologies and base-point placements. (15)

UNION OF SOVIET SOCIALIST REPUBLICS

The 370km EEZs of the Soviet Union and the United States overlap in three areas - the Chukchi Sea, the Bering Sea, and the North Pacific Ocean. The USSR-US maritime boundary is the longest oceanic boundary in the world - 3330 km (1800 nm) and was established in the 1867 United States/Russia Convention ceding Alaska to the US (Figure 11.4). Not until the establishment of the FCMA was it important to accurately define the limits of these

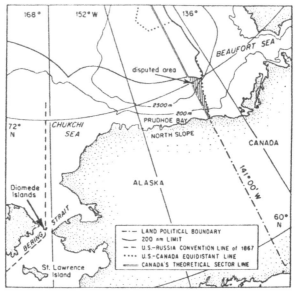

Figure 11.3. United States-Canada Beaufort Sea disputed area. After W.G. Gordon, and R.E. Gutting Jr, _Oceanus_, _27_, 4: (Winter 1984/85), p.42

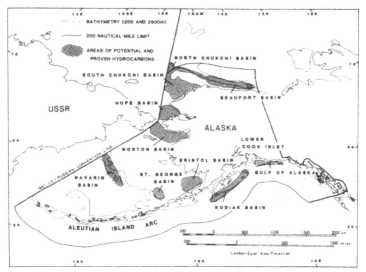

Figure 11.4. United States-Soviet Union EEZ boundaries and basins with gas and oil potential. After R.W. Rowland _et al._ p.9, (see Figure 11.1)

waters; now, with the EEZ and with oil exploration leasing under way in the region, precise boundary delimitation is imperative.

Several incidents related to fishing in the Bering Sea have caused US Government officials to think that the Soviets might be calculating the position of the boundary differently. This assumption proved true, perhaps because no chart accompanied the 1867 Convention. In 1981, when EEZ boundary talks began, Soviet negotiators told their US counterparts that they were using a rhumb line (a straight line on a Mercator projection) to delimit the boundary; the US was using arcs of great circles (the shortest distance and a straight line on a globe). This methodological difference creates a wedge-shaped area that each state considers as under its control. As of early 1986, after five rounds, negotiations continue, (16) but their successful resolution may be difficult, given the region's potential for hydrocarbons. On the other hand, some observers suggest that the two states jointly should exploit petroleum resources in the disputed portion of the Navarin Basin. Cooperation might be patterned after the UK and the Netherlands' North Sea arrangements in their jointly held oil and gas fields, whereby costs and proceeds are divided based on proportional field ownership. If US-USSR cooperation cannot be effected, perhaps a 'buffer zone' can be established in which neither state can explore or exploit non-living resources until a boundary line is delimited. (17) In fact, the US already has moved in this direction.

Because of the undelimited boundaries, the Department of the Interior (DOI) and the Department of State prohibit drilling or exploitation in the disputed area. The DOI, however, has made one lease sale (March 1983) in the Navarin Basin; its leasing procedures for the disputed waters are carefully designed. These procedures provide that the bid of the highest bidder on any given tract offered for sale which lies within the disputed zone is put into escrow. No bids will be accepted or leases given unless the US Government decides that it is in its best interest. If, after five years, the bid has not been accepted by the Government, the highest bidder may withdraw bid money by giving notice within 60 days after the lapse of the five years. (18)

MEXICO

The US and Mexico share three maritime boundaries. Seven months before the US implemented its FCMA in 1977, Mexico declared a 370km EEZ; boundary negotiations between the US and Mexico began almost immediately, resulting in 1976 in provisional lines. These boundaries were formalised in 1978 by the Treaty of Maritime Boundaries, but the US Senate has not yet ratified it. The treaty, based on equidistance and taking islands into full account, includes boundaries in both the Pacific Ocean and the Gulf of Mexico. Because the Parties agreed to deal only with those areas within 370km of both their coasts, the Gulf of Mexico has an undelimited 239km (129nm) segment (Figure 11.5). (19)

THE CARIBBEAN

The US and Cuba have a common boundary for 556km (300nm). It extends from the Gulf of Mexico to a point off Florida's south coast which is equidistant from the US, Cuba, and the Bahamas. Upon

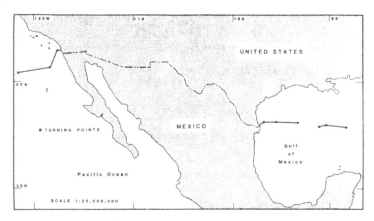

Figure 11.5. United States-Mexico maritime EEZ boundaries. Source: After NACOA, (see note 11, page 40)

the establishment of the FCMA in 1977, the Cuban and US Governments entered into discussions which led to the development of two equidistant lines - one using a straight baseline system (adhered to by Cuba), and

another using the low-water line system (adhered to
by the US). The coastal configuration of northern
Cuba is not deeply indented or fringed by islands;
thus, using a straight baseline, in this instance,
does not meet the criteria of international law as
prescribed in Article 4 of the 1958 Geneva
Convention on the Territorial Sea and the Contiguous
Zone. Nevertheless, on 16 December 1977, the two
states reached agreement on a compromise boundary
line that divides the disputed area equally but not
equidistantly. In early 1978, the treaty reached the
US Senate and in 1980 its Foreign Relations
Committee favourably reported on the treaty. As of
early 1986, however, the Senate had not acted to
ratify, mainly because of continuing poor relations
between Havana and Washington, DC. (20) The 1977
agreement remains in force, being reviewed every two
years. (21)

When the US established its 370km fishery zone
in 1977, the Bahamas was informed of the location of
the provisional enforcement boundary; because the
distance between the two land areas is less than
740km, the boundary was drawn using the equidistance
method. Ths US and the Bahamas have never initiated
negotiations on this boundary. The DOI plans to hold
a petroleum lease sale near the boundary area in
1986, but it will not offer tracts in the overlapping zone. (22)

On 20 November 1980, the US and Venezuela
established a treaty boundary that lies south of
Puerto Rico, extending for approximately 370km. The
boundary's western terminus is not yet delimited,
because negotiations with the Dominican Republic are
not completed. The terminus, however, will not lie
west of longitude 68°51'28"W and latitude
15°14'28"N. The Dominican Republic and the US have
overlapping EEZs off Puerto Rico. Inconclusive
discussions begun in 1977 were abandoned, then
reinstituted in 1983. Negotiations have been
amicable but have not been concluded. (23)

Between their Virgin Island territories, the
United Kingdom and the US have established a
Reciprocal Fisheries Agreement boundary based on the
equidistant method. The two governments have not
initiated discussions on the boundary between
Anguilla and the US Virgin Islands. Neither have
there been formal negotiations between the
Netherlands and the US concerning waters between the
US Virgin Islands and the Netherlands Antilles. (24)

THE PACIFIC

A treaty with New Zealand on 3 September 1983 established a boundary between American Samoa and Tokelau, settled conflicting claims to three islands in the Tokelauan group, and confirmed the US' sovereignty over Swains Island. On 8 September 1983, the Cook Islands and the US established a treaty of friendship, settling disputes over four islands and delimiting a boundary between American Samoa and the Cook Islands. Both the New Zealand and Cook Islands treaties also settled fishery disputes. As of early 1986, no boundary negotiations for waters between the United States' Jarvis, Howland, and Baker Islands and the Republic of Kiribati have been initiated. (25) A treaty of friendship, however, went into force 23 September 1983, resolving the ownership of several small islands and providing nondiscriminatory rights of access for fishermen of Kiribati to supply canneries in American Samoa. A boundary remains to be negotiated with Japan between Iwo Jima and the Northern Mariana Islands where overlapping EEZs occur. Boundaries between Western Samoa, Tonga, the Marshall Islands, and the Federated States of Micronesia must also be established. (26)

THE EEZ AND MINERALS

The US Government is moving quickly to evaluate the resource potential of its EEZ. For example, the National Oceanographic and Atmospheric Agency (NOAA), in anticipation of a multiple-use of the EEZ, is developing a series of four atlases at a scale of 1:4 000 000. The atlases focus upon the Bering, Chukchi, and Beaufort Seas; the Gulf Coast of Alaska and the coasts of Washington, Oregon, and California; the East Coast; and the Gulf of Mexico. Each atlas contains data '... on the space-time distribution of selected characteristics of each region, including: 1) physical environments; 2) biotic environments; 3) living marine resources; 4) economic activities; 5) environmental quality; and 6) political boundaries and jurisdictions.' (27) Presently, the United States Geological Survey (USGS) is developing a Continental Margin Map (CONMAP) at a scale of 1:1 000 000 that will show a synthesised and currently available data base for the '... structure, sedimentary framework, and stratigraphy of the EEZ.' The USGS has also mapped

some 647 500km² (250 000mi²) of the sea floor in the
EEZ off the coast of California, using a sidescan
sonar system developed in the UK. (28) The USGS also
has identified, for detailed study, 12 major
corridors within the EEZ that represent diverse
geological conditions and provide a broad
representation of the continental margins. Seismic
lines and both deep and shallow-water drilling have
been undertaken in these corridors (Figure
11.6). (29)

In addition to hydrocarbons, there are five
major categories of hard minerals in numerous areas
of the EEZ which, in most cases, have been neither
fully evaluated for their production potential nor
extensively exploited. These categories include
construction materials, placers, phosphorites, polymetallic sulfides, and cobalt-ferromanganese crusts.
Sand and gravel, calcareous shells, sulfur, salt
brines, and barite have been dredged or pumped in
near-shore areas of the US. In future, these
important chemical and construction minerals (30)
should continue to be important in the United
States' offshore mineral economy, and both placers
and phosphorites will become increasingly so. (31)

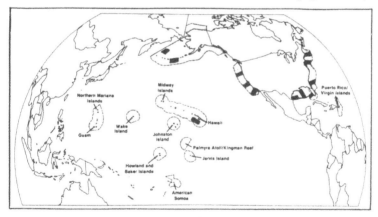

Figure 11.6. Corridors representative of diverse
geological conditions in the EEZ. Source: D.L. Peck,
p.81, (see note 29)

Immediate interests, however, have been stimulated
most by the recently discovered polymetallic
sulfides and cobalt-ferromanganese crusts.

POLYMETALLIC SULFIDES

The polymetallic sulfides are associated with sea-floor vulcanism, especially in sea-floor spreading zones. Discoveries of hydrothermal venting along the East Pacific Rise near the southern end of the Gulf of California at 21°N led to further exploration along the West Coast of North America. Scientists have found several vents 'smokers' in the southern part of the Juan de Fuca Ridge spreading zone, just outside the United States' EEZ. These hydrothermal vents emit plumes of hot, mineral-rich waters that react with the cold sea water to form sulfide and oxide precipitates of lead, copper, zinc, and iron, among others. The nearby Gorda Ridge spreading zone, lying within the EEZ, off the coast of northern California and southern Oregon may contain similar vents (Figure 11.7).

Although the extent of polymetallic sulfide deposits in the Gorda Ridge area is undetermined, the DOI's Minerals Management Service (MMS) has pushed forward rapidly to make the area available for leasing. Many critics say the MMS has moved too quickly and in an effort to preempt jurisdiction from NOAA in the Department of Commerce (DOC). (32) During 1983 the MMS drafted an environmental impact statement and in 1984 held environmental impact hearings in Oregon and California. The main issues raised by hearings participants concerned the potential impacts of mining on biological organisms associated with the vents, on water quality, on fisheries, and on coastal settlements. (33) Because of these uncertainties, the DOI - along with other federal agencies and the Governors of California and Oregon - formed a task force to advise the DOI on the feasibility of proceeding with the lease sale. As of early 1986, the task force had not completed its work. (34) Whether those in industry will respond favourably when the lease sale is held is in doubt; investors are fearful because mining legislation assuring them an affordable access (as depletion allowances and credits for exploration costs) and specifying the 'rules of the game' has not been developed by the MMS and The Congress. (35) Indeed, many in industry say they will 'wait and see' what happens both in The Congress and in world metal markets. With present world metal prices, with continuing jurisdictional ambiguities between the DOI and DOC, and with a still inadequately developed seabed mining technology for extracting polymetallic sulfides, industry is unlikely to follow the eager lead of the federal government.

COBALT-FERROMANGANESE CRUSTS

A more recent discovery than the polymetallic sulfides is extensive cobalt-ferromanganese crusts found coating the sides of volcanic islands and some 100 known seamounts in the United States Pacific islands EEZs, especially in waters off the Hawaiian and Line Islands. The US must import nearly all its cobalt from areas, such as Zaire, that are or have been politically unstable. Cobalt crusts are richest at depths of 1000-2500m (3280-8200ft) and contain up to 2.5 per cent cobalt, with concentrations of 16kg (35.3lbs) of ore per m^2 (10.8ft^2). The crusts also contain nickel, copper, molybdenum, and platinum, as well as manganese - all minerals with strategic value and appeal to industry and the US Government. In fact, the economic potential of these crusts is greater than the world's prime manganese nodule deposits that cover the Pacific basin's deep seabed between the Hawaiian Islands and Mexico. (36) Unfortunately, the crusts' main location - on the sides of often steep and irregularly shaped seamounts - will make their extraction very expensive, even if the mining technology were available, and it is not.

CONCLUSIONS

What does the future hold for the US EEZ? Given that the 370km EEZ concept has been part of customary international law for a decade, to debate the United States' right to unilaterally declare an EEZ seems irrelevant. With the EEZ in place and the national government working to identify its mineral potential, there are increased incentives to establish formal boundaries with states having contiguous and overlapping EEZs. Considering its past successes, the US should succeed in peacefully and equitably delimiting these boundaries.

Although the US Government hopes industry will proceed 'full speed ahead' to exploit minerals of the EEZ, as in the Gorda Ridge area, industry's response may be unenthusiastic, perhaps until well after the turn of the century. Until federal legislation precisely defines the rules for offshore hard-minerals producers, until more detailed geological data are available on the various offshore minerals, and until more attractive metal markets

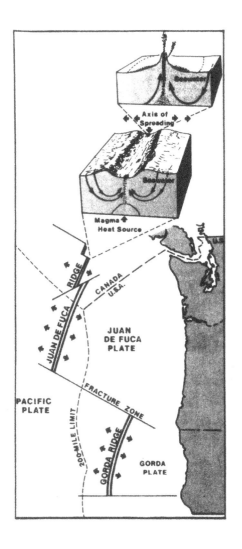

Figure 11.7. Juan de Fuca Ridge and Gorda Ridge spreading zones. Source: M.A. Champ, W.P. Dillon, and D.G. Howell, 'Non-living EEZ resources: minerals, oil and gas', <u>Oceanus</u> <u>27</u>, 4: (Winter 1984/85), p.32

develop, industry may choose to keep its speculative ventures onshore.

NOTES

1. R. Reagan, 'Proclamation 5030 of March 10, 1983: Exclusive Economic Zone of the United States of America', Federal Register, (14 March 1983), 48, 50, 10605.

2. W.P. Pendley, 'Importance of the EEZ Proclamation', in Symposium Proceedings: A National Program for the Assessment and Development of the Mineral Resources of the United States Exclusive Economic Zone, Nov. 15, 16, 17, 1983. USGS Circular 929 (Alexandria, Va, 1984), 5.

3. H.S. Truman, Policy of the United States with Respect to the Natural Resources of the Subsoil and Sea Bed of the Continental Shelf, Proclamation 2667, Sept. 28, 1945, as cited in Public Papers of the Presidents of the United States: Harry S. Truman, (Washington DC, 1961), 352-3.

4. G.A.B. Pierce, 'Selective adoption of the new law of the sea: the United States proclaims its exclusive economic zone', Virginia Journal of International Law, 23, 4, (1984).

5. United Nations, The Law of the Sea: United Nations Convention on the Law of the Sea (New York, 1983), 18.

6. Ibid, 27-8.

7. Ibid, 22-3.

8. Ibid, 24-5.

9. Ibid, 21-2.

10. G.A.B. Pierce, op.cit, (1984), 595-6.

11. National Advisory Committee on Oceans and Atmosphere, The Exclusive Zone of the United States: Some Immediate Policy Issues, (Washington DC, 1984),

12. Ibid, 38.

13. 'International Court of Justice case concerning delimitation of the maritime boundary in the Gulf of Maine area (Canada/United States of America) Oct. 12, 1984', International Legal Materials, 23: Nov. 1984, 246-398.

14. R.W. Smith, Letter: Special Assistant for Ocean Affairs and Policy Planning, Office of The Geographer, Department of State (Washington DC), Letter, 13 Jan., 1986.

15. H.R. Marshall, 'Disputed areas influence OCS leasing policy', Offshore 45, 5: (1985) 99-100.

16. Ibid, 100-101.

17. K.M. Shusterich, 'Arctic issues coming to fore', Oceanus 27, (1984-85) 4:82.

18. H.R. Marshall, op.cit., (1985) 100-101.

19. Ibid, 101.

20. R.W. Smith, loc.cit. (1986), H.R. Marshall, op.cit., (1985) 102.

21. D.C. Slade, 'Maritime boundaries of the United States', in NACOA, op.cit., (1984) (note 11), 41.

22. H.R. Marshall, op.cit., (1985) 41, 45.

24. R.W. Smith, loc. cit., (1986).

25. R.W. Smith, loc. cit., (1986).

26. D.C. Slade, op.cit., (1984), 45-6.

27. B.A. McGregor, and M. Lockwood, Mapping and Research in the Exclusive Economic Zone (Reston, Va., 1985) 36.

28. Ibid, 8, 16.

29. D.L. Peck, 'The US Geological Survey program and plans in the EEZ', op.cit., (1984, note 2), 80.

30. F.C.F. Earney, (1980) Petroleum and Hard Minerals from the Sea (Arnold, London, 1980), 11-26.

31. J.R. Moore, 'Alternative sources of strategic minerals from the seabed', in G.J. Mangone, (ed.), American Strategic Minerals (New York, 1984) 85-108.

32. J.M. Broadus, and P. Hoagland III, 'Rivalry and coordination in marine hard minerals regulation', in National Oceanic and Atmospheric Administration, Exclusive Economic Zone Papers, (Rockville, Md, 1984), 55-61.

33. Draft Environmental Impact Statement: Proposed Outer Continental Shelf Polymetallic Sulfide Minerals Lease Offering, Gorda Ridge Area Offshore Oregon and Northern California, (Minerals Management Service, DOI, Reston, Va, 1983), i-ii.

34. K. Davidson, Staff Attorney, NOAA, DOC, (Reston, Va., Letter, 15 Jan. 1986).

35. J.R. Moore, 'OSIM for Marine Mining', Marine Mining, 5, 3:(1986), 335-6.

36. W.P. Dillon, and D.G. Howell, 'Nonliving EEZ resources: minerals, oil and gas', Oceanus 27, 4: (1984-1985), 31-2.

ACKNOWLEDGEMENT

Figures 11.3 and 11.7 are reproduced by kind permission of Oceanus, Woods Hole, Ma.

CHAPTER 12

HISTORICAL GEOGRAPHY AND THE CANADA-UNITED STATES
SEAWARD BOUNDARY ON GEORGES BANK

Louis De Vorsey

> My boundaries enclose a pleasant land;
> indeed, I have a goodly heritage.
>
> Psalm 16:6

> Mr President, I wish to make clear at the
> outset what it is that brings the Parties
> before the Chamber on this occasion. In two
> words, it is Georges Bank. The written
> pleadings of both Parties leave no room for
> doubt that the object of their dispute is
> Georges Bank.
>
> Mark McGuigan,
> Minister of Justice and Attorney
> General of Canada

> Dividing Georges Bank between us would create
> difficult problems of joint fishery con-
> servation and management, problems that would
> persist and fester.
>
> Davis R. Robinson,
> Legal Advisor,
> United States Department of State

On October 12, 1984, a five-member Chamber of
the International Court of Justice at the Hague
delivered its judgement in the case concerning the
delimitation of the United States - Canadian lateral
maritime boundary in the Gulf of Maine and seaward
over the rich Georges Bank fishing grounds. (1) The
parties had brought their dispute for arbitration
under the terms of a Special Agreement signed in
1979. (2) In that agreement they requested the Court
to decide the course of a single maritime boundary
that divided their continental shelf and fisheries
zones from a point off the coast of Maine designated

Canada-United States Boundary on Georges Bank

A and a large triangular-shaped area in the Atlantic Ocean seaward of Georges Bank (Figure 12.1). Discussions concerning the delimitation of a continental shelf boundary were held between the US and Canada, as long ago as 1975, but no agreement could be reached. In 1977, the implementation of 200-mile fishery zones by both countries created pressure for a speedy resolution of the problem as it touched the Gulf of Maine and Georges Bank fishing grounds. Lateral boundary extension seaward to the 200 mile limit was required through the disputed area of overlapping national claims.

In September, 1978, an announcement in the Canadian Gazette further expanded Canada's claim in the region. The United States responded by asserting that the new Canadian claim had no merit under international law and would not be recognised. Finally, in 1979, the treaty containing the agreement alluded to above was signed and ratified.

INTERNATIONAL COURT OF JUSTICE CHAMBER ESTABLISHED

Early in 1982, the International Court of Justice ordered a Chamber of five judges constituted to hear the dispute. Four judges in the International Court were selected to sit on the Gulf of Maine Chamber: Andre Gros of France; Hermann Mosler of West Germany; Roberto Ago of Italy; and Stephen Schwebel of the United States. Maxwell Cohen, former Dean of McGill University, Faculty of Law, was appointed by Canada to sit on the Chamber as judge ad hoc. (3)

August 26, 1982, was fixed as the date for the submission of Memorials by both Canada and the United States. The Memorials are massive printed documents running to six volumes in the case of the United States and five volumes for the Canadian argument. No expense was spared in collecting and presenting the evidence to support each side's contentions. Many of the coloured maps and illustrations included would look at home in glossy publications such as the Canadian Geographical or National Geographic Magazine. (4)

After six months, during which experts on each side analysed the other's arguments and supporting evidence, a second or Counter-Memorial was filed with the World Court. The Counter-Memorials were delivered for exchange on June 28th, 1983. As in the case of the Memorials, they are lavishly illustrated multi-volume printed documents. (5) Needless to say,

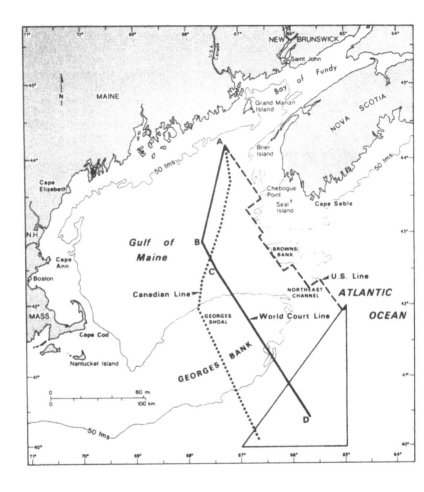

Figure 12.1. The United States' and Canadian claims in the Gulf of Maine, with the World Court line

the Counter-Memorials were carefully analysed by the opposing parties. After several months of hard work, responses to the Counter-Memorials were prepared. These responses, usually referred to as Reply Briefs or simply Replies, were submitted on December 12, 1983, and in the Spring of 1984, the World Court Chamber heard oral arguments from the representatives of the United States and Canada. (6) Because of an interruption occasioned by Nicaragua's case against the United States, the oral arguments could not be completed as originally planned. The arguments commenced on April 2nd and were finally concluded on May 11th of 1984.

THE DISPUTE AS SEEN BY THE UNITED STATES AND CANADA

Before turning to the Chamber's decision, it is instructive to gain some perspective on the general nature and significance of the case as seen by the parties. In Canada's view:

> ... the range of factors in issue gives this case a unique human immediacy and economic importance. For the first time, the Court is asked to direct its attention to the full panoply of sovereign rights and jurisdiction that coastal States may now exercise in the area beyond and adjacent to the territorial sea with respect to the living and non-living natural resources of the seabed and subsoil and the superadjacent waters. The case thus brings together the promise of the continental shelf and the known abundance of the fishery: the economic prospects that the shelf may offer in years to come and the present livelihood of the coastal fishermen and the welfare of the communities they sustain. (7)

In his opening oral presentation outlining the United States point of view for the Chamber, Davis R. Robinson, Legal Advisor, United States Department of State stressed that:

> In the United States, the intense concern with this case stems largely from historical American links with Georges Bank. United States fishermen have fished significantly on Georges Bank since the 1820s. A rich folklore developed surrounding the exploits and the daring of these New Englanders. For almost a

century and a half, it was with few isolated exceptions, United States vessels alone that fished the waters over Georges Bank. During this period, especially in New England, the fisheries of Georges Bank were, to use a rather common American metaphor, considered by many citizens to be as "American as apple pie". In brief, Georges Bank has been closely connected with the United States for a long time. As could be expected, Canada's aspirations regarding the northeast portion of Georges Bank provoked a strong response, not only in New England, but also in the corridors of Washington. (8)

THE DELIMITATION LINES PROPOSED BY THE PARTIES

In their pleadings both Canada and the United States presented specific lines which connected point A with the area of the large triangle described in their Special Agreement. These proposed delimitation lines are shown on Figure 12.1. Not surprisingly, the one proposed by the United States quickly earned the sobriquet, 'the wedding cake line'. On the other hand, although not precisely a line composed of points which were equidistant from the adjacent coasts, the Canadian line approached that quality and came to be referred to as 'the equidistant line' or 'Canadian line'.

In the opinion of the Chamber neither of these boundary lines had merit. 'In both cases', the majority of judges agreed, 'the outcome of the Parties' efforts can be said to have been pre-conceived assertions rather than any convincing demonstration of the existence of the rules that each had hoped to find established by international law.' (9) Clearly the Chamber viewed the voluminous collections of legal and factual findings submitted by the parties flawed by the 'a priori nature' of their underlying premises and deductions.

The Chamber was further critical of the parties' 'false premise' which led them to search and employ international law 'for, as it were, a set of rules which were not there'. (10) 'Customary international law', the Chamber averred, 'comprises a limited set of norms for ensuring the co-existence and vital cooperation of the members of the inter-national community'. There does, in the Chamber's opinion, exist a set of 'customary rules' whose presence 'can be tested by induction based on the analysis of a sufficiently extensive and convincing practice, and not by deduction from preconceived ideas.' (11) A body of detailed rules for such a delimitation is not what the Chamber would accept as a bonafide product resulting from a search of the extant relevant body of international law.

EQUITABLE CRITERIA AND PRACTICAL METHODS

In moving toward its determination the Chamber reached the conclusion that: 'a delimitation by a single line ... ie a delimitation which is to apply at one and the same time to the continental shelf and to the superadjacent water column can only be carried out by the application of a criterion, or combination of criteria, which does not give preferential treatment to one ... to the detriment of the other, and at the same time is such as to be equally suitable to the division of either of them.' (12)

The criteria which the Chamber ultimately chose were derived from geography. By geography, however, the Chamber meant 'mainly the geography of coasts, which has primarily a physical aspect, to which may be added, in the second place, a political aspect.' (13) Within such a framework the Chamber favoured 'a criterion long held to be as equitable as it is simple, namely that in principle ... one should aim at an equal division of areas where the maritime projections of the coasts of the States ... converge and overlap.' (14) The geographical complexities of the New England - Nova Scotia facade fronting the Gulf of Maine would, of course, make necessary 'the likewise auxiliary criterion whereby it is held equitable partially to correct any effect of applying the basic criterion that would result in cutting off one coastline, or part of it, from its appropriate projection across the maritime expanses to be divided, or then again the criterion - it too being of an auxiliary nature - involving the necessity of granting some effect, however limited, to the presence of a geographical feature such as an island or group of small islands lying off a coast, when strict application of the basic criterion might entail giving them full effect or, alternatively, no effect.' (15)

Once committed to this line of reasoning the Chamber found that geometrical methods would be suitable instruments for giving effect to their criteria for delimitation. In the words of the

Decision, 'The delimitation line to be drawn in a given area will depend on the coastal con-figuration.' (16) Considered as a whole, the Chamber's approach to the problem of deciding a single delimitation line for the continental shelf and waters above it subscribed to the adage in law that 'the land dominates the sea'. (17)

CHAMBER'S CONSTRUCTION OF THE DELIMITATION LINE

To begin the construction of a delimitation line the Chamber established what it considered to be the coastal areas of the United States and Canada fronting on the Gulf of Maine. In the Chamber's view the Gulf of Maine formed 'a broad oceanic indentation in the eastern coast of the North American con-tinent, having roughly the shape of an elongated rectangle.' (18) In their construct neither the waters of the Bay of Fundy nor those of Massachusetts Bay and Cape Cod Bay were included in the Gulf of Maine. The approximately rectangular geometry of the Gulf of Maine as defined by the Chamber can be observed on the map (Figure 12.2).

Beginning at the terminus of the US-Canada international boundary, the Chamber found that a straight imaginary line drawn through Grand Manan Island to Brier Island and Cape Sable would form the eastern side of a rectangle. A bit over two hundred miles to the southwest a similar imaginary line drawn between 'the elbow of Cape Cod' and Cape Ann formed the opposite and 'quasi-parallel' side of the Chamber's rectangle. The long sides were formed by the seaward closing line between nantucket Island and Cape Sable and another imaginary line connecting Cape Elizabeth, Maine and Grand Manan Channel. This approximate rectangle then became the framework within which the Chamber employed its criteria and methods to determine an equitable line of delimitation from point A through the area of overlapping Canadian and US offshore claims.

In brief, they drew perpendiculars from Point A to the two imaginary coastal rectangle lines which formed a corner at the terminus of the international boundary. They then determined the reflex angle formed at Point A. This angle was approximately 278°. (19) The bisector of this angle was taken as the course or azimuth for the first segment of the delimitation line in the Gulf of Maine.

To determine the course of the second segment the Chamber looked again at the geography of the

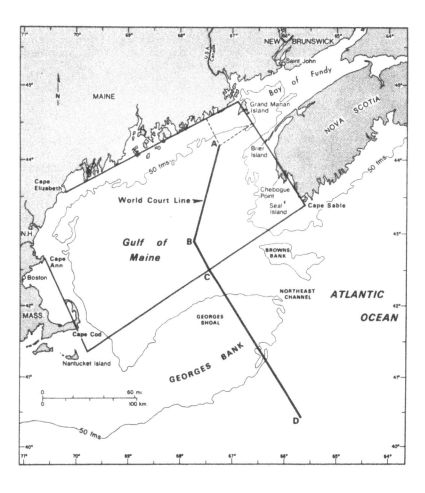

Figure 12.2. The World Court definition of the coasts of the United States and Canada fronting the Gulf of Maine

Gulf of Maine as they had defined it. In this review they noted that a distinctly different relationship existed between the coasts of Nova Scotia and Massachusetts than the one which existed between Nova Scotia and Maine. Massachusetts and Nova Scotia were found to have opposite coasts and did not have the quality of lateral adjacency which had guided their determination of the first segment of the line from Point A. Here, the Chamber stated, only a median line drawn between the quasi-parallel lines of the opposing coasts would satisfy their requirement for a geometrical method. A correction was needed, however, to take account of the difference in the lengths of the US and Canadian coastlines. In the Chamber's reckoning there were about 284 miles of US coastline as compared to 206 miles of Canadian involved. This produced the ratio of 1.38 to 1. A further correction was deemed necessary, however, before an equitable or adjusted median line could be fixed upon. (20)

The Chamber considered that Seal Island and Mud Island, off the coast of Nova Scotia, should have some influence in the calculation of the location of the second segment of the delimitation line. Rather than transfer the whole of the Nova Scotian coastline by the distance between Seal Island and the mainland coast the Chamber decided to allow the island half effect. This resulted in a corrected ratio of 1.32 to 1 in place of the 1.38 to 1 which was determined before Seal Island was taken into account. In the Chamber's words 'the result of the effect to be given to the island is a small transverse displacement of the (corrected median) line, not an angular displacement; and its practical impact therefore is limited. (21)

Thus the second segment of the delimitation line was the corrected median line between the coasts of Nova Scotia and that of Massachusetts as the Chamber had defined them. It intersected the first segment at Point B on the map (Figure 12.2) and extended seaward to the closing line of the Gulf of Maine. The Chamber noted 'that the meeting-point of the first and second segments of the delimitation line, ie the pivotal point where this line changes direction, is located about as far into the Gulf as Chebogue Point, a feature of the Nova Scotian coast which marks the transition from the part of this coast in an adjacency relationship with the coast of Maine to the part facing the Massachusetts coast in a relationship of oppositeness.' (22)

The third and longest segment of the delimitation line would lie entirely seaward of the Gulf of Maine closing line. As a consequence of this fact, the Chamber reasoned that the most appropriate geometrical method for its determination would be the simple drawing of a perpendicular from the closing line at the point of intersection with the corrected median line. This point is shown as C on Figure 12.2. From Point C the Chamber's line of delimitation was extended across Georges Bank into the triangular area the Parties had designated in their Special Agreement. There the line of delimitation was terminated where this perpendicular from the Gulf of Maine closing line reached the outermost extent of the overlapping 200 mile claims of Canada and the United States. While it is probably incorrect to view the Chamber's efforts in terms of winning or losing, one cannot help but be impressed by the fact that its line of delimitation confirmed to Canada a large portion of Georges Bank. In the words of the Chamber's decision 'this Bank is the real subject of the dispute ... the principal stake in the proceedings, from the viewpoint of the potential resources of the subsoil and also, in particular, that of fisheries that are of major economic importance.' (23) Clearly the geography to geometry methodology followed by the Chamber, in their effort to articulate the 'fundamental norm of customary international law governing maritime delimitation', had resulted in a division of Georges Bank not radically different from the 'equidistance' division which had been argued by Canada.

EQUITABLENESS OF CHAMBER'S LINE

It was not until nearly the end of the printed Decision that the Chamber turned to, 'Its last remaining task' which was 'to ascertain whether the result thus arrived at may be considered as intrinsically equitable, in the light of all the circumstances which may be taken into account for the purpose of that decision.' (24) In the instance of the first two segments of the delimitation line within the Gulf, the Chamber did not consider such verification to be 'absolutely necessary'. Outside of the Gulf and over Georges Bank, the Chamber stated that the question might take on a 'different complexion'. The Chamber saw Georges Bank as 'the real subject of the dispute between the United States and Canada in the present case'. The Bank and the fishery it supported were without any doubt 'the principal stake in the proceedings' the Chamber was hearing.

The Chamber then posed for itself the rhetorical question as to whether or not there might be factors beyond those 'provided by the geography of the Gulf itself' which should have been taken into account in assessing the equitable character of the delimitation line over Georges Bank. (25) It was unfortunate that the Chamber waited until almost the end of its adjudicatory process to finally undertake some consideration of this truly remarkable submarine feature dominating the disputed oceanic area. Perhaps, given the Chamber's already noted espousal of the 'land dominates sea' perspective in the case, such an outcome was predictable. By adopting this perspective however, the Chamber was left with what Judge Gros termed 'a sea deprived of all meaning, an empty sea, which is to be divided.' (26) It was as if the continuous light blue colouring, which cartographers frequently employ to represent oceans on their maps, had been accepted as a reasonable reflection of historical and physical reality. Nothing could be further from the true dynamic reality of Georges Bank.

HISTORICAL GEOGRAPHY AND THE CONCEPT OF THE BEHAVIOURAL ENVIRONMENT

To gain a fuller appreciation of the reality of Georges Bank the Chamber needed only to broaden its narrow definition of geography sufficiently to embrace the sub-discipline of historical geography. Historical geographers would approach Georges Bank as an enormously significant field of human activity, and recognise that an understanding of those activities depends on the skilful reconstruction of both the area's past phenomenal environment and past behavioural environment.

In his article 'Historical Geography and the Concept of the Behavioural Environment' British geographer, William Kirk, pointed out that: 'In as much as in Historical Geography we are concerned with the behaviour of human groups in relation to environment it behoves us to reconstruct the environment not only as it was at various dates but as it was observed and thought to be, for it is in this behavioural environment that physical features acquire values and potentialities which attract or repel human action.' (27) Kirk's behavioural environment will be recognised as similar to the

'psychomilieu' discussed by Harold and Margaret
Sprout in the context of international political
decision-making. In the Sprouts' view, 'what matters
in decision making ... is how the milieu appears to
the decision makers under consideration - not how
the milieu actually is or how it appears, or might
appear, to some other persons.' (28)

On the other hand, the phenomenal environment,
in Kirk's view, 'is an expansion of the normal
concept of environment not only to include natural
phenomena but environments altered and in some cases
almost entirely created by man.' (29) Kirk's schema
provides a succcinct explanation of his concepts;

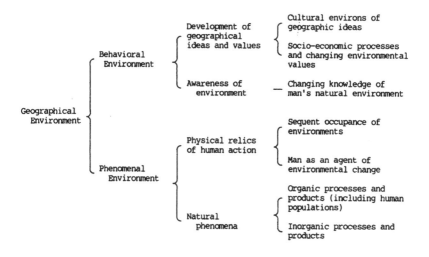

THE OFFSHORE ENVIRONMENT AS CONSIDERED BY THE CHAMBER

It is significant to note that neither of the
Parties invoked a consideration of the historical
behavioural environment of Georges Bank in its
presentations to the Chamber. The United States
mounted historical arguments which were summarised
in the following paragraph of the Chamber's
Decision:

233. In the eyes of the United States, the
main consideration here is the historical
presence of man in the disputed areas. It
believes the decisive factor here to be the

activities pursued by the United States and its nationals since the country's independence and even before, activities which they claim to have been alone in pursuing over the greater part of that long period. This reasoning is simple and somewhat akin to the invocation of historic rights, though that expression has not been used. This continuous human presence took the form especially of fishing, and of the conservation and management of fisheries, but it also included other maritime activities concerning navigational assistance, rescue, research defense, etc. All these activities, said greatly to exceed in duration and scale the more recent and limited activities of Canada and its nationals, must, according to the United States, be regarded as a major relevant circumstance for the purpose of reaching an equitable solution to the delimitation problem. (30)

Canada, as might be expected, placed far less stress on the historical role of Georges Bank. In the Chamber's words 'the only period which in Canada's eyes should be regarded as relevant was the recent one leading up to, or even continuing beyond, the time when both States finally decided to go ahead with the institution of exclusive fishery zones.' (31)

The Chamber rejected these positions of the Parties. In the case of the United States, the Chamber confirmed 'its decision not to ascribe any decisive weight, ... , to the antiquity or continuity of fishing activities carried on in the past within that part of the delimitation area which lies outside the closing line of the Gulf.' (32) In the Chamber's view the waters over Georges Bank 'were part of the high seas and as such freely open to the fishermen not only of the United States and Canada but also of other countries.' (33) More discussion of the United States' thesis regarding Georges Bank is included in the Chamber's Decision but these quotations suffice to make clear that the Judges persisted in a view of these disputed waters as what Judge Gros critically characterised as 'an empty sea' - just so much more blue 'high seas' as shown on the maps and charts.(34)

Most of the 300 supporting maps, sketches or diagrams which Canada and the United States included with the some 7,600 pages of pleadings and 2,000

pages of oral arguments submitted to the Chamber
showed the bathymetry and features of Georges Bank.
It is a dominant submarine feature of shallows
immediately seaward of the Gulf of Maine closing
line. In anthropomorphic terms it might be described
as a broad submerged 'thumb' stretching from a
'hand' or 'palm' formed by the northeastern states.
The space between the imagined thumb and an index
finger, forming the New England coast from
Massachusetts to New Brunswick, is occupied by the
Gulf of Maine. Immediately to the north of the tip
of the Georges Bank 'thumb' lies a marked depression
shown on the charts as the Northeast Channel.
   After its consideration of the best and most
current expert scientific evidence dealing with the
offshore environment, which the Parties could supply
the Chamber concluded that:

> The continental shelf of the whole of this
> area is no more than an undifferentiated part
> of the continental shelf of the eastern
> seaboard of North America, from Newfoundland
> to Florida. According to generally accepted
> scientific findings, this shelf is a single
> continuous, uniform and uninterrupted physio-
> graphical structure, even if here and there it
> features some secondary characteristics
> resulting mainly from glacial and fluvial
> action. In this wider context the continental
> shelf of the area relevant to the present
> proceedings may be defined as the natural
> prolongation of the land mass around the Gulf
> of Maine; neither Party disputes the fact that
> there is nothing in this single sea-bed,
> lacking any marked elevations or depressions,
> to distinguish one part that might be
> considered as constituting the natural pro-
> longation of the coasts of the United States
> from another part which could be regarded as
> the natural prolongation of the coasts of
> Canada. (35)

Even the disputed area's most accentuated feature,
the Northeast Channel was found to 'lack the
characteristics of a real trough marking the
dividing-line between two geomorphically distinct
units.' (36)
   When it came to the water over and around
Georges Bank, the Chamber concluded 'that it too
essentially possesses the same character of unity
and uniformity already apparent from an examination

of the sea-bed, so that, in respect of the waters too, one must take note of the impossibility of discerning any natural boundary capable of serving as a basis for carrying out a delimitation of the kind requested of the Chamber.' (37) Neither Party, needless to say, had argued the existence of a 'natural boundary' in the waters of the disputed area seaward of the Gulf of Maine. They had, however, emphasised the distinctive and palpable qualities of those waters at great length. Canada, for example, acknowledged the fact that, 'there is a distinct ecosystem on Georges Bank, which is geographically defined by the Great South Channel and the Northeast Channel.' (38)

It is clear from these quotations, and other statements by the Chamber, that it was not swayed from its 'empty sea' stance by the Parties' most impressive efforts to employ the findings of modern ocean science concerning the 'Phenomenal Environment' of the area in dispute. Modern science can inventory and evaluate the natural phenomena, living and inanimate, as well as the physical and human processes operating in any particular area of earth space with a high degree of reliability. Oceanographic scientists do not, however, normally concern themselves with the reconstruction of past 'Behavioral Environments' - those cognition fields or milieus within which individuals and groups framed their choices, decisions and activities. It is suggested that this is precisely the sort of reconstruction with which the Chamber should be primarily concerned if the application of international law is to result in truly equitable solutions rather than apparently arbitrary decisions concerning the human disputes and problems which result from those past choices, decisions and actions.

THE BEHAVIOURAL ENVIRONMENT OF GEORGES BANK

The full reconstruction of the Behavioural Environment of Georges Bank and its disputed waters would run to an extended length and produce a monograph of ample proportions. Space limitations will not permit the examination of more than a few selected elements which could form integral parts of the reconstructed Behavioural Environment of the disputed area.

In the eyes of the many generations of fishermen who inhabited the shores of the Gulf of Maine and began the commercial exploitation of its fishery at least as early as 1710, Georges Bank was an extension of New England. This view was eloquently stated by the famous New England philosopher-author, Henry David Thoreau in the middle of the last century. In his book <u>Cape Cod</u> Thoreau reported that 'On Cape Cod the next most eastern land you hear of is St George's Bank'. Thoreau went on to state:

the fishermen tell of 'Georges', 'Cashus', and other <u>sunken lands</u> they frequent. Every Cape man has a theory about George's Bank having been an island once, and in their accounts they gradually reduce the shallowness from six, five, four, two fathoms, to somebody's confident assertion that he has seen a mackerel-gull sitting on a piece of dry land there. (emphasis added) (39)

The persistence of this folk view of Georges Bank as a former island outlier of New England is demonstrated by Shebnah Rich in an account published in 1883;

The distance from Chatham to southwest Georges is about sixty miles. One sixth nearly of this distance is here accounted for by islands which had been eroded by the sea since Gosnold and Smith's exploration in the early 1600s, which well sustains the opinion, that at no very distant day Georges Bank was connected with groups of islands, if not mainland, extending to Nantucket and the Cape. Old Skipper Joseph Wharf, the father of the late Joseph, used to say that he had played ball on Georges, and men were living fifty years ago, who said they had seen long strings of gulls sitting on the dry sandbars. It used to be quoted as history, that an Amsterdam Company once proposed purchasing the right to build there a port. (40)

Henry Mitchell, a scientist with the US Coast and Geodetic Survey, added qualified support to folk cognitions of Georges Bank as a wasted remnant of fomer land. In an official report dated October 10, 1878, Mitchell noted that 'St George's Bank, the summit of which is called George's Shoal, is probably a wasted island.' In support of this conclusion Mitchell wrote:

Capt. Isaac Hull, USN, made note, in 1815(?), of the assertion of Cape Cod fishermen that 'part of the shoal has been quite dry, with gulls sitting upon it,' and states that his own shoalest water was three feet after subtracting over seven feet for height of tide. There would seem to be no reason why the sands should not heap up occasionally so as to emerge at low tide; but the testimony is not direct, and I find much the same report mentioned, with discredit in Hollingsworth's Nova Scotia, printed in 1786. (41)

It is informative to contrast these reported folk views of Georges Bank as a wasted island with the analysis of Thomas Pownall who served as governor of colonial Massachusetts during the period 1757-1759. In his frequently cited treatise, Topographical Description of the Dominions of the United States of America, Pownall described 'a long Hook of a promontory called Cape Cod.' Pownall continued:

The long Low neck of land by which this (the Promontory itself is high Land) is joined to the Main seems to have been formed by the Coil and Recoil of the tides, rolling up Silt and Sand at the Thread of their least Force. ... Many and various Alternations have been made, and are continually making on the East Coast at the Back of this Promontory and a long point of sand has been formed into solid Marsh Land within these Forty years, at the South Point of it. Let those who are curious in the Process of the Operation of Nature, watch the Progress of George's Sand. From the Inquiries I made, and the Answers I got, I think that it will in some years, and perhaps not many hence, form into another Sable Island. Its south Point is now at Low Water with a strong off-Shore Wind visibly a shoal. (42)

For maximum appreciation, Pownall's description of Georges Bank in process of evolving into another Sable Island should be read with a copy of 'A General Map of the Northern British Colonies In America ..., 1776' at hand. (43) On this important Revolutionary-era map, a large shaded area off the New England coast is identified as 'St George's Bank or Malabar'. Several small detached shaded areas fringe the main bank on its southern and south-eastern edge. A gloss on the map explains these areas as 'South Points of Malabar to be dry at low water with an offshore wind vid. Governor Pownall's Topographical Description.' Thus Pownall's hypothesis concerning Georges Bank, based on 'Inquiries I made, and the Answers I got', was brought to the attention of a wider map-using public as well as the readers of his book.

Pownall was not alone in his interest in the formation and dynamics of the fishing banks found on the continental shelves of New England and eastern Canada. Thomas Jefferson, while considering a proposal to construct a canal through the Isthmus of Panama in a leter written in 1786, commented on possible impacts on the fishing banks far to the north. He speculated that the Gulf Stream would be deflected by such a canal and 'Those banks, ceasing to receive supplies of sand, weeds and warm water by the gulph stream, it might become problematical what effect changes of pasture and temperature would have on the fisheries.' (44) Jefferson was not alone in attributing delivery of the sands forming the offshore fishing banks to the north-flowing Gulf Stream.

Even more interesting than his speculation regarding the influence of the Gulf Stream in forming and nourishing the fishing banks were Jefferson's thoughts concerning the ocean current as a potential 'natural boundary' for the United States. John Quincy Adams revealed Jefferson's ideas on this score in the following extract dated November 30, 1805:

> The President mentioned a late act of hostility committed by a French privateer near Charleston, S.C., and said that we ought to assume as a principle that the neutrality of our territory should extend to the Gulf Stream, which was a natural boundary, and within which we ought not to suffer any hostility to be committed. ... I observed that it might be as well, before we ventured to assume a claim so broad, to wait for a time when we should have a force competent to maintain it. But in the meantime, he said, it was advisable to squint at it, and to accustom the nations of Europe to the idea that we should claim it in future. (45)

While the United States never made an official attempt to establish its outer limits at the Gulf

Stream, there is evidence that much 'squinting' to seaward did take place in the period following Jefferson's statement. For example, in 1807, 'An Act to Provide for Surveying the Coasts of the United States' declared it to be lawful for the President 'To cause such examinations and observations to be made, with respect to St Georges Bank, and any other bank or shoal ... to the Gulf Stream ...' (46) The specific inclusion of 'St Georges Bank' in this general statute is an indication of just how highly regarded this portion of continental shelf was at this early date. From the fishing villages around the Gulf of Maine to the corridors of power in the new capital city of Washington, Georges Bank conjured up images of both potential riches and great hazard for the hardy ship owners and fishermen who dared venture there.

The unpredictable weather, wave conditions, and swift tidal currents which swept the shallow seas over Georges Bank were responsible for an enormous loss of human life and property. The port of Gloucester, Massachusetts, alone counted the loss of 87 ships and 722 men on Georges Bank in the period from 1837 to 1873. (47) Not surprisingly this factor loomed large in the emergence of Georges Bank as a major element of the Behavioural Environments of the groups inhabiting the shores of the Gulf of Maine. In the words of an early Chairman of the Portland Harbor Commission 'the bones of all who have perished there would make a monument higher than that on Bunker Hill; and the property lost would pay for one of silver, if not too high.' (48)

The surviving brother of one victim of Georges Bank was moved sufficiently to develop a truly amazing scheme to convert the shoalest portion of the bank to a man-made island to serve as the base for a lighthouse. In 1838, Seward Porter, a well known Portland shipowner, petitioned Congress 'for the erection of a beacon on St George's shoal, or the cession of said shoal to him and his associates,' (49) (emphasis added). Porter proposed that 'old vessels loaded, with stone, to be sunk and imbedded there' would form the nucleus of an island much like Sable Island to the north. In his view 'the Isle of Sables was once a hidden shoal, and converted into an island by the accidental wreck of vessels thereon.'

Nothing came of Porter's proposal to gain a cession of Georges Shoal from Congress, but the idea of creating an island there persisted. Almost half a century later, Henry Mitchell, Assistant of the US

Coast and Geodetic Survey presented 'an urgent plea for the establishment of a light on Saint George's Bank' as Appendix No.11 to the Superintendent's annual report for 1885. Mitchell like Porter drew attention to the similarity which existed between Georges Shoal and Sable Island in the following statement from his plea:

> Saint George's Bank lies at the threshold of the Gulf of Maine, and its summit is called distinctively 'George's Shoal' or 'The Georges'. Twenty miles westward of the Georges is the 'Cultivator Shoal'. Upon the Georges there are two spots of 12 feet given by chart, while upon Cultivator Shoal there is not less than 18 feet. ... It is 100 miles from the nearest land, Chatham, Cape Cod. Its situation relative to the continent is like that of Sable Island, and like this island, it would long ago have been lighted if <u>dry land</u>. As a shoal it is far more dangerous than it would be as an island, and therefore it is shunned and feared except by fishermen who are obliged to brave its danger. (50)

In both implicit and explicit terms, Porter's scheme and Mitchell's later plea for a lighthouse on Georges make clear the fact that the bank was widely recognised as being within the domain of the United States. Clearly the early cognitions of the common fishermen, who saw Georges Bank as 'the next most eastern land' from their New England shores, were being confirmed in these ambitious proposals to convert the shoal to an island for a light house. Such cognitions, when taken in the aggregate, outline the historical Behavioural Environment of Georges Bank and form the basis for special circumstances which were eminently worthy of consideration by the International Court of Justice in its attempt to structure an equitable line of delimitation.

Thus far in this discussion the significance of Georges Bank in the Behavioural Environments of New Englanders and national leaders of the United States has been stressed. It is necessary, in view of the dispute and decision under consideration, to direct some attention to Canadian historical cognitions with respect to Georges Bank. When Canada's stress on the recent period is recalled it should come as no surprise that even the most diligent search of the historical record failed to produce a significant body of Canadian cognitions which could be used to argue that Georges Bank was historically a true Canadian Behavioural Environment. What that search did reveal was substance for an argument against any such claim had Canada attempted to mount one.

One extremely convincing body of evidence which would negate an attempt to place Georges Bank in a position of prominence in any reconstructed Canadian Behavioural Environment is found in the records of the Halifax Fisheries Commission, an international tribunal which convened for several months during 1877. (51) The Commission was charged with determining what, if any, compensation the United States should pay Canada in return for the 'exchange' of inshore fishing rights under the provisions of the 1870 Treaty of Washington. To support their claims the Canadians collected depositions from several hundred fishermen. Each of the 319 deponents described the areas in which he had fished in his career at sea. Several of them mentioned seasons during which they worked aboard American vessels, with some mentioning time spent on George's Bank or in the 'George's Fleet'. In all of this voluminous testimony, however, only one Canadian vessel is described as ever having fished on George's Bank. Julien Boudreau, of Esquimaux Point, Quebec, testified:

I have lived here 16 years, before that I lived at the Magdalen Islands where I was born. I am 63 years of age and have been a fisherman for 50 years, and for the last 45 years I have been carrying on the fishery with a vessel of which I was the master and owner, on the north coast of the Gulf of St Lawrence, from Sheldrake to the Straits of Belleisle, and on the Atlantic Coast of Labrador, as far as Cape Harrison, at the Magdalen Islands, on La Havre Bank, at the mouth of the Bay of Fundy, and on Georges Bank. I am well acquainted with every operation relating to the taking and curing of codfish, halibut, mackerel and herring. (52)

The best that can be concluded is that, at some time between 1843 and 1878, Julien Boudreau, a Quebec fisherman and master, fished on Georges Bank. This single instance out of 319 investigations does little to place Georges Bank high in a Canadian Behavioural Environment or domain of any sort.

The absence of Canadian fishing vessels on Georges Bank in this nineteenth-century period is also verified by a survey of the annual lists of shipwrecks published by the Canadian Department of Marine and Fisheries. These lists normally include the location at which vessels were wrecked or lost. A screening of the lists for 1869, 1870, 1871, and 1872 indicates no wrecks or losses in the area of Georges Bank. (53)

Similar negative evidence for the period 1884-1890 can be gleaned from a history of Yarmouth, Nova Scotia, one of the Canadian fishing centres closest to Georges Bank. This work includes a comprehensive list of Yarmouth County maritime casualties during this period. (54) What emerges from a study of these data is the fact that most of the Yarmouth ships lost were engaged in merchant shipping rather than fishing. The fishing vessels which were lost met their fates on the Grand Banks, Western Bank, or Grand Manan. No Yarmouth vessels were lost on Georges Bank although several Yarmouth fishermen lost their lives there while serving on New England schooners out of Gloucester and Rockland.

CONCLUSION

In its effort to establish an equitable line of delimitation through the disputed waters off of New England and Nova Scotia, the International Court of Justice paid close attention to the physical and political geography of the coasts framing the Gulf of Maine. Based on its interpretation of that complex geography, the Court adopted a geometrical methodology to draw a line whose segments depended on particular coastal configurations around the Gulf. By extending this methodology seaward of the Gulf's closing line the Court awarded approximately half the disputed area of Georges Bank to each contesting party, Canada and the United States. In the Court's view this approximately equal division of the area of overlapping claims to the Atlantic's richest fishing ground was equitable.

In so doing the Court disregarded the historic positions of Canada and the United States in the disputed continental shelf area known as Georges Bank and treated it as 'an empty sea', in the words of the dissenting Judge. Had the Court been made aware of the role of Georges Bank as a Behavioural Environment through its long history of human use a rather different outcome, one more favourable to the

United States, might have resulted.

NOTES

1. See, International Legal Materials, XXIII, No.6, Nov., 1984, 1197-1269.

2. For text of the Special Agreement, see International Legal Materials, XXIII, No.6, 1200-1202. (The geographic coordinates of Point A and the triangular area are given on p.1201.)

3. Ibid, 1200.

4. International Court of Justice, Delimitation of The Maritime Boundary In The Gulf of Maine Area (Canada/United States of America) Memorial Submitted by Canada, September, 1982; and International Court of Justice Case Concerning Delimitation Of The Maritime Boundary In The Gulf of Maine Area (Canada/United States of America) Memorial Submitted By The United States of America, 27 September, 1982.

5. International Court of Justice, Delimitation Of The Maritime Boundary In The Gulf of Maine Area (Canada/United States of America) Counter-Memorial Submitted by Canada, 28 June, 1983; International Court of Justice, Case Concerning Delimitation Of The Maritime Boundary In The Gulf of Maine Area (Canada/United States of America) Counter-Memorial Submitted By The United States of America, 28 June, 1983.

6. International Court of Justice, Delimitation Of The Maritime Boundary In The Gulf of Maine Area (Canada/United States of America) Reply Submitted By Canada, 12 December 1983; International Court of Justice, Case Concerning Delimitation Of The Maritime Boundary In The Gulf of Maine Area (Canada/United States of America) Reply Submitted By The United States of America, 12 December 1983.

7. Canada, Memorial, 18.

8. David R. Robinson, Verbatim Record, Case Concerning the Delimitation of the Maritime Boundary in the Gulf of Maine Area, 11 April 1984, p.7-8.

9. International Legal Materials, XXIII, No.6, 1223.

10. Ibid.

11. Ibid, 1224.

12. Ibid, 1238.

13. Ibid.

14. Ibid.

15. Ibid, 1238.

16. Ibid, 1239.

17. (Re - Land Dominates Sea).

18. I.L.M. XXIII, No.6, 1208.

19. Ibid, 1241.

20. Ibid, 1242.

21. Ibid, 1243.

22. Ibid.

23. Ibid, 1244.

24. Ibid.

25. Ibid.

26. Ibid, 1259.

27. William Kirk, 'Historical Geography and the Concept of the Behavioural Environment', Indian Geographical Society Silver Jubilee and N. Subrahmanyam Memorial Volume (Madras, 1952), 159; and his subsequent article, 'Problems of Geography', Geography, 48 (1963), 357-71.

28. Harold Sprout and Margaret Sprout, Toward A Politics of the Planet Earth (New York: 1971), 192.

29. W. Kirk, op.cit. (1963), 364.

30. I.L.M. XXIII, No.6, 1244-45.

31. Ibid, 1245.

32. Ibid.

33. Ibid.

34. Ibid.

35. Ibid, 1211.

36. Ibid.

37. Ibid, 1213.

38. Ibid, 1212.

39. Henry David Thoreau, Cape Cod, Vol.2 (Boston: 1896), 123.

40. Shebnah Rich, Truro-Cape Cod or Land Marks and Sea Marks, (Boston: 1883), 195.

41. Quoted by Henry Mitchell in 'Physical Hydrography of the Gulf of Maine', Appendix No.10, US Coast and Geodetic Survey, Report of the Superintendent for 1879 (Washington; 1879), 176.

42. Mitchell's statement that the report of dry banks on Georges Bank was mentioned with 'Discredit' by Hollingsworth is hard to understand and may have resulted from an incorrect reading of his notes. With reference to Georges Bank Hollingsworth wrote, 'it is asserted to have been seen dry in some places, which is not improbable as there are credible persons who have sounded upon it in three fathoms water.' See, S. Hollingsworth, The Present State of Nova Scotia: With A Brief Account of Canada ..., 2nd ed. (Edinburgh; 1787), 29.

43. Robert Sayer and John Bennett, A General Map of the Northern British Colonies in America ... (London; 1776).

44. Julian P. Boyd (ed.), The Papers of Thomas Jefferson, Vol.10, (Princeton; 1954), 530.

45. John Quincy Adams, Memoirs, Vol.1, ed. by Charles Francis Adams (Philadelphia; 1874), 375.

46. US Laws, Statutes (1841), p.78.

47. George H. Proctor, The Fisherman's Memorial and Record Book (Gloucester, 1873), 9-53.

48. Quoted by Henry Mitchell in 'A Plea for a Light on Saint George's Bank', Appendix 11, in US Coast and Geodetic Survey, Report of the Superintendent for 1885 (Washington: 1886), 485.

49. Seward Porter, 'Memorial for the Erection of a Beacon on St Georges Shoal, or the Cession of Said Shoal to Him and His Associates', in Senate Documene 115, 25th Congress 2nd Session (Washington: 1838), 1.

50. H. Mitchell, see note 48 above, p.483.

51. Award of the Fishery Commission. Documents and Proceedings of the Halifax Commission, 1877, Under the Treaty of Washington of May 8, 1871. 3 Vols (Washington: 1878).

52. Ibid, Appendix G, 178-80.

53. See Parliamentary Sessional, Papers of the Dominion of Canada for 1870, 1871, 1872, and 1873, Annual Reports of the Department of Marine and Fisheries (Ottawa; 1870, 1871, 1872, and 1873).

54. J. Murray Lawson, Yarmouth Past and Present: A Book of Reminiscences (Yarmouth; 1902).

CHAPTER 13

MARITIME BOUNDARIES IN THE MEDITERRANEAN: ASPECTS OF
COOPERATION AND DISPUTE

Nurit Kliot

The Mediterranean Sea is a semi-enclosed sea
with a common heritage of navigation and commerce.
The Mediterranean basin is not a homogeneous region:
the eighteen Mediterranean states belong to three
continents and many of their cultural, political and
economic networks lie within these three continents.
There is great ethnic, cultural and religious
diversity among Mediterranean states and the
economic gap between poor and rich countries around
the Mediterranean basin is very wide. (1) This
cultural and economic contrast has led to
differential interests and varied policies toward
the sea and its resources. Moreover, complex coastal
configurations and the presence of islands create
geographical and legal problems for boundary
delimitation. (2) The task of drawing maritime
boundaries in the Mediterranean has also been made
difficult by British ownership of Gibraltar and the
British bases of Cyprus, and by the Spanish
ownership of Ceuta and Melilla in North Africa.
There are longstanding disputes over maritime and
territorial issues involving Greece and Turkey,
Spain and the United Kingdom, Spain and Morocco and
Israel and its Arab neighbours. Thus, both geographical and political circumstances make agreement
on maritime boundaries difficult.
States which have the necessary strength to
defend their claims, may claim rights over the
adjacent seas, for a variety of purposes, and for
differing distances. For coastal states there are
five commonly recognised zones: internal waters,
territorial waters or seas, contiguous zones, and
continental shelf. The need to draw an international
maritime boundary arises and the process of boundary
formation normally begins when two states make
claims to waters or parts of the seabed which

overlap. Agreement on the boundary may be achieved through unilateral or bilateral action and sometimes the parties will need arbitration.

This paper will examine patterns of cooperation and conflict, of agreements and disagreements on maritime boundaries in the Mediterranean Sea. Specifically, it will focus on national claims to maritime jurisdictions in the Mediterranean, on methods of delimitation of maritime boundaries in the Mediterranean Sea and on agreements and disputes concerning maritime boundaries.

THE 1982 LAW OF THE SEA: NEW CONCEPTS

The third UN Conference on the Law of the Sea (UNCLOS) which was concluded in 1982 with a new convention, has dramatically changed the maritime legal regime in the Mediterranean. The most prominent changes are four:
1) Every Mediterranean state has the right to extend the limits of its territorial sea to 12 nautical miles (nm).
2) The contiguous zones may be increased from the former 12 nm to 24 nm.
3) The new concept of Exclusive Economic Zone of 200 nm from the baseline of the territorial sea, if adopted in the Mediterranean, will subject the whole sea to the jurisdiction of the coastal states.
4) The 1982 Convention makes considerable changes in the right of free transit in international straits. A new concept of 'transit passage' has been introduced which cannot be suspended by the coastal state and applies also to aircraft. The states bordering straits may adopt laws and regulations relating to transit passage, concerning the safety of navigation and maritime traffic, the prevention, reduction and unloading of any commodity or persons according to the customs, fiscal, immigration to sanitary laws and regulations of coastal states.

THE PRACTICE OF THE 1982 CONVENTION AMONG MEDITERRANEAN STATES: TERRITORIAL WATERS

Table 13.1 summarises the practices of the 1982 Convention in the Mediterranean Sea. The adoption of the 1982 Convention had highlighted many of the current disputes and conflicts over maritime jurisdiction. The table shows that fourteen of the eighteen Mediterranean states claim a territorial sea of 12 nm. Greece and Turkey claim a territorial sea of 6 miles (Turkey only in the Aegean) because any expansion of the territorial sea to 12 nm in this area would turn the Aegean Sea into a Greek 'lake' and might turn a politically dangerous dispute to an overt conflict. (3) Israel claims only 6 nm in the Mediterranean because of its sensitive position in the Red Sea which the Arab states have tried to turn into an 'Arab lake' while Israel has tried to preserve its right of free navigation in this sea. This may at least partly explain Israel's rejection of the 1982 convention. (4) The other reason for Israel's general objection to UNCLOS Convention was its reference to non-government groups such as the PLO in seabed mining provisions.

The extension of Albania's territorial sea to 15nm could create problems with its neighbours mainly concerning fishing rights, and the extension of the Syrian territorial sea to 35nm may lead to a dispute with Turkey. The off-shore boundary between these two states has not been delimited and is not agreed. Table 13.1 shows that these two countries did not sign the 1982 Convention probably because of their exceptional territorial waters claims.

THE MEASUREMENT AND DELIMITATION OF MARITIME BOUNDARIES IN THEORY AND PRACTICE

One of the most problematic issues concerning maritime boundaries in the Mediterranean and else-where is the process of boundary delimitation. According to Prescott (1985) there are three types of circumstances which make negotiation, measure-ment and delimitation of maritime boundary a difficult task. First, political circumstances such as a state of war, lack of formal relations or poor relations between states, will probably prevent the start of negotiations on delimitation. Second, geographical circumstances might make negotiations and agreement difficult. There are problems associated with maritime boundaries of islands, whose location may be very inconvenient for one of the parties. There is also the problem of the nature of the seabed and the question of natural prolongation. Thus, countries which are restricted by alien islands or which have only short coasts, or which are closely surrounded by neighbours, are likely to argue strongly against the use of equidistant lines as maritime boundaries. Third, there are economic circumstances which might

Table 13.1: National Claims to Maritime Jurisdiction in the
Mediterranean to 1985

Countries	Territorial Sea(nautical) miles	Continental Shelf(*)	Year Straight Baselines proclaimed	1982 Convention signed
Albania	15 (1976)	Unspecified	1960	–
Algeria	12 (1972)	Nil	1964	1982
Cyprus	12 (1964)	E	–	1982
Egypt	12 (1958)	200m or E	1951	1982
France	12 (1971)	Unspecified	1967	1982
Greece	6 (1936)	200m or E	–	1982
Israel	6 (1964)	E	–	–
Italy	12 (1974)	200m or E	1978	1984
Lebanon	12 (1983)	Nil	–	1984
Libya	12 (1959)	Unspecified	1973	1984
Malta	12 (1978)	200m or E	1972	1982
Monaco	12 (1973)	Nil	–	1982
Morocco	12 (1973)	200m or E	1975	1982
Spain	12 (1977)	Unspecified	1977	1982
Syria	35 (1981)	200m or E	1963	–
Tunisia	12 (1973)	Unspecified	1973	1982
Turkey	6 Aegean Sea (1964)	Unspecified	1964	–
	12 Black Sea (1973)			
	12 Mediter- ranean (1973)	–		
Yugoslavia	12 (1979)	200m or E	1965	1982

*   E = claim to depth of exploitation.
   200m = claim to 200 metres depth

Sources: J.R.V. Prescott, The Maritime Political Boundaries of
the World, (Methuen, London: 1985): R. Smith (ed.) National
Claims to Maritime Jurisdictions, Department of State, Limits
in the Seas, 36, 5th Rev. (Washington, March 1985).

complicate negotiations and agreements on boundaries. If there is a marked disparity in wealth and resources between two states, the poorer state may argue for a larger share of the disputed zone. Also, agreement might be made more difficult if the economic potential of a disputed area is high. (5)

The theory and practice of maritime boundary delimitation and measurement will be analyzed under the headings of 'straight baselines proclamation' and 'continental shelf delimitation'.

STRAIGHT BASELINE PROCLAMATION

The 1982 Convention (Article 5) states that the normal baseline for the measurement of maritime zones is the low-water mark along the coast. Straight baselines can be used to replace the low-water mark as the baseline in two situations. First, when there are local circumstances where short straight lines are appopriate: they include the mouths of rivers and small bays. Since the maximum closing line for a bay is 24nm, these local straight baselines will generally be shorter than that distance. Second, where the coast is highly indented or fringed with islands, the use of longer straight baselines may be appropriate.

The Conventions of 1958 and 1982 also allow state jurisdiction over 'historic' bays. This rule permitted states to escape from the provisions concerning the drawing of closing lines and defining legal bays. This escape is simplified by the lack of codification of international law regarding historic bays. (6)

Table 13.1 shows that 13 Mediterranean states have proclaimed straight baselines. Albania estab-lished straight baselines in 1960. These are considered by some to be inappropriate along Albania's uncomplicated coast. Egypt and Syria established seven inappropriate straight baselines. Their main departures from the Convention permit shoals to be used as turning points, allow bays to be closed without any reference to width or depth and authorises the inclusion of any islands. Neither have produced charts showing their straight base-lines. Egypt has claimed the Bay of el Arab as an historic bay. (7)

France established straight baselines in 1967. The use of straight baselines here seems appropriate. Italy defined its baselines in 1978. According to Prescott, there are several sections in its straight baselines which are open to criticism. Italy also claims the Gulf of Taranto as an historic bay. (8)

Libya did not establish straight baselines but in 1973 claimed the Gulf of Sirte (or Sidra) as Libyan internal waters. The closing line measures approximately 300nm. This claim was justified on the grounds that Libya had exercised sovereignty over the waters throughout history, and on the grounds of national security. As soon as the Libyan claim was made, objections were raised by Britain, France, Italy, Greece, Russia, Turkey and the United States. The USA has conducted periodical naval exercises in the vicinity of the Gulf since 1977. In 1981 and 1985-1986 the exercises led to serious disputes between the two states. According to Prescott there is no evidence that Libya has exercised exclusive sovereignty over the claimed waters, and plenty of evidence to show that Libyan claims to that effect have been challenged by other countries. (9)

Malta proclaimed straight baselines in 1972. Its inclusion of the island of Filfla in the Maltese baseline is perhaps legally unjustified. (10) Spain defined straight baselines in the Mediterranean Sea and the Atlantic Ocean in 1977. With a few exceptions the Spanish coast is too smooth and lacking in islands to justify the use of straight baselines. In 1973, Tunisia proclaimed straight baselines and proclaimed the Gulf of Gabes (which is too large to be a legal bay) as an historic bay. Turkey announced its extensive straight baselines in 1964, which are generally considered to be appropriate. Yugoslavia's straight baselines (1965) could serve as a model of modest baselines resulting in minimal claims to internal waters.

The improper use of straight baselines has not, so far, provoked any serious international disagreement or conflict. The conflict that erupted over the Gulf of Sirte was concerned with the closing line at 32°30'N and not the baseline.

CONTINENTAL SHELF DELIMITATION IN THE MEDITERRANEAN

The applicable principle for measuring the territorial sea is the median line. As for exclusive economic zones and continental shelf boundaries, their formation should be effected by Article 38 of the statute of the International Court of Justice, which stresses an equitable solution. This new emphasis follows judgements of the International

Court of Justice which found that the rule of equidistance and the rule relating to special circumstances had the same aim, which is the delimitation of a boundary in accordance with equitable principles.

A careful examination of continental shelf agreements shows that the different views expressed above are reflected in the practice of boundary delimitation. The Agreements on the continental shelf boundaries between Italy and Yugoslavia and Italy and Tunisia are based on an adjusted median line, taking into account islands and islets. (11) An equidistance line was developed in the Italian-Tunisian agreement, but four Italian islands of Lampedusa Lampione, Linosa and Pentelleria which lie on the Tunisian side of the line, were given territorial seas of 12nm or 13nm. (12) The Agreement between Greece and Italy was based on a median line with significant adjustment near the Greek islands of Othoni and Kefallinia. (13) The continental shelf boundary between Italy and Spain also used equi-distance. The continental shelf boundary between Italy and Yugoslavia was based on the equidistance principle, but the Italians were compensated for the Yugoslav islands which are located well out in the Adriatic Sea. Altogether, the Yugoslav concessions to the Italians amounted to 3080sq.km, 1680sq.km in the north and 1400sq.km in the south, while the Italian concession to Yugoslavia was 416sq.km. (14)

The Agreement between France and Spain was based mainly on a negotiated line, only a minor part of which was the median line. (15) Because France supports equitable principles while Spain and Italy support the median principle their respective maritime boundaries reflect both principles.

Perhaps one of the most important inputs to the delimitation of continental shelf boundaries in the Mediterranean has been provided by the International Court of Justice decisions, particularly in the Libya-Tunisia case. The Court ruling of 1982 was highly controversial and has been severely criticised for the lack of method in the Court's approach. (16) According to the 1958 and 1982 Conventions, the basic principle in delimitation of the continental shelf when states fail to agree is one of equidistance. But in the Tunisia-Libya case, the Court stated that equidistance is neither a mandatory legal principle nor a method having some privileged status. The court decided on a maritime boundary line of a bearing of 26° to the meridian, and then 52°, choosing this line partly because it reflected the manner in which both Libya and Tunisia granted concessions for off-shore exploration and exploitation. (17) The dispute between Libya and Malta on their continental shelf boundary was also referred to the Court in July 1982. This dispute became a conflict when in August 1980 a Libyan frigate prevented work on one of the oil rigs with which Malta was exploring the Medina Bank. (18) In July 1984, the International Court of Justice ruled on a line between Malta and Libya. Malta had wanted the boundary drawn on the principle of equidistance, but the Court ruled that because of Libya's longer coastline opposite the island, the line should be set 18nm further north. This is currently under appeal. (19)

Italy and Malta are currently engaged in negotiating their continental boundary in an area which has become a busy oil and gas search region. Malta and Tunisia have also to settle their continental shelf boundary. In fact, the whole area between Tunisia, Libya, Malta, and Italy is disputed among these states. This region is of major economic importance as rich oil and gas sources have been found in it (see Figure 13.1).

AGREEMENTS AND DISAGREEMENTS ON MARITIME BOUNDARIES

A maritime boundary may be agreed through unilateral agreement, bilateral agreement or arbitration. Unilateral action in relation to overlapping maritime claims means that one country proclaims a maritime boundary of, for example, 15nm without any active confrontation from its neighbours, who refuse to be provoked when their nationals are prevented from fishing or exploring the seabed in the zone claimed by this state. If one state has taken a passive role while the other state defended its claim then a de facto boundary would result, favouring the defensive state. If both states adopted defensive attitudes, a conflict would arise. According to Prescott (1985) maritime boundaries have not generally resulted from the assertion of force. (20)

Bilateral action takes place when the parties discuss and negotiate a boundary line. If the parties are unable to reach agreement, the governments involved may leave it unresolved or refer it to a tribunal such as the International Court of Justice or to arbitration. (21)

Table 13.2 and Figure 13.1 present the agreed

Table 13.2: Agreement and Disagreement on Maritime Boundaries in the
Mediterranean 1985

Boundary	Unilateral declarations (a)	Bilateral agreement (b)	Arbitration by I.C.J.	Disagreement
T.W. = Territorial waters			C.S. = Continental Shelf	
Spain - Morocco				C.S/enclaves
Spain - France		working arrangement on C.S.		
Spain - Italy		C.S. 1978		
Tunisia - Libya			C.S. 1982	
Tunisia - Italy		C.S. 1978		
Italy - France		working arrangement on C.S.		baselines in the North Tyrrhenian Sea
Italy - Malta		working arrangement on C.S.		
Italy - Yugoslavia		C.S. 1970		
Italy - Albania	T.W.	working arrangement on C.S.		
Italy - Greece		C.S. 1980		
Italy - Libya				Gulf of Sirte closing line
Greece - Albania	T.W.			
Greece - Turkey				Aegean Sea C.S.
Greece - Libya				Gulf of Sirte closing line
Turkey - Cyprus		working arrangement on C.S.		
Syria - Turkey	T.W.			
Libya - Malta			C.S. 1985	
Britain - Cyprus		T.W. around bases		

Sources: A.D. Drysdale and G.H. Blake (1985), see note 28.
S.P Jagota, Maritime Boundary. (Martinus Nijhoff, The Hague 1985)
Annex 1.
G. Luciani, (ed.) The Mediterranean Region - Economic Interdependence
and the Future of Society, (Croom Helm, Beckenham, 1984).
J.R.V. Prescott, The Maritime Political Boundaries of the World.
(Methuen, London, 1985) 299-304.

Notes (a) Unilateral action is defined as one-side announcement of
outer-limits which extend beyond the 1982 convention
12.n.m limits.
(b) Bilateral Agreements include 'working arrangements'
between neighbouring countries short of formal
agreement.

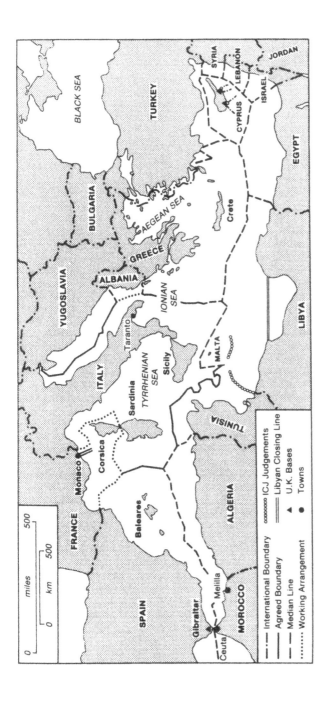

Figure 13.1 Maritime boundaries in the Mediterranean Sea

and disputed maritime boundaries in the
Mediterranean. Under 'Unilateral agreements' two
maritime boundaries are represented with a question
mark, because the unilateral announcements of both
Albania and Syria on outer limits of their
territorial sea of 15nm and 35nm respectively were
objected to by several states. (22)

The unilateral action of Libya in closing the
Gulf of Sirte was objected to firmly by at least
five Mediterranean countries and therefore cannot be
defined as an agreement.

The bilateral agreements in Table 13.2 do not
include the Black Sea. Italy is a partner to four
continental shelf agreements with Spain, Tunisia,
Yugoslavia and Greece. It is interesting to note
that Italy has not attained any boundary agreement
with France. Spain, too, has not reached agreement
with France though they each have an agreement on
their continental shelf boundary in the Atlantic
Ocean. It seems that the continental shelf boundary
between France and Italy has not been agreed because
of the existence of islands, while they are in
dispute over the definition of baselines in the
northern Tyrrhenian Sea, between Corsica and
Tuscany. Disagreement between Italy and France is
probably related to the fact that the two countries
disagree on a method of delimitation: France favours
the equitable principle while Spain (and Italy)
support the median or equidistance principle.

There is also an agreement between Britain and
Cyprus which specifies territorial waters around
British bases in Cyprus. The boundaries as shown on
Figure 13.1 do not have any seaward terminus and
therefore they can accommodate the British claim of
3 nautical miles and the Cypriot claim of 12
nautical miles. Altogether there are 7 maritime
boundaries agreed among 18 neighbouring states, –
approximately 14% of the potential boundaries. But
there are more disputes than agreements in the
Mediterranean and some of these have evolved into
high tension, as in the Aegean Sea. The dispute
between Turkey and Greece relates to the presence of
more than 1,000 Greek islands from Samothrace in the
north to Rhodes in the south. It is what Blake
(1986) calls a positional dispute, involving overlapping claims in a complex geographical area. (23)

Turkey consistently maintains that the area in
which the Greek islands are situated is part of the
Turkish continental shelf and that the Greek islands
do not possess a continental shelf of their own. The
Greek claim is that in the absence of an agreement

between the two states and of any special circumstances justifying another boundary line, the
delimitation must be based upon the median line
between the Greek islands and the Turkish
coast. (24) In 1976 both countries threatened to use
naval force over the despatch of a Turkish seismic
vessel to disputed waters in the Aegean sea. Greece
mobilised its Aegean Fleet but the conflict was
contained after the intervention of NATO to which
both countries belong. In March 1984, the Greek
Government complained that five Turkish destroyers
had fired salvoes at a Greek destroyer off
Samothrace. (25) This incident was also restrained
but as there are no negotiations between the two
states on their maritime problems, this dispute can
be expected to flare up again. The conflict is
exacerbated by indications of oil in the Northern
Aegean sea. Greece requested the International Court
of Justice to determine as to whether or not certain
islands under Greek sovereignty are entitled to a
continental shelf of their own, but the Court
decided that it was without jurisdiction to
entertain the Greek application. (26)

The other disagreements in the Mediterranean
fall into two classes. First, there are other
problems of boundary delimitation in which the
presence of islands deflects lines of equidistance
to the disadvantage of neighbouring countries. There
is a group of Spanish islands, Islas Chaferinas,
situated 4nm from Cap de Agua on the Moroccan coast.
Their presence results in the theoretical median
line between Morocco and Spain being marked by a
deep protuberance in favour of Spain (Figure 13.1)
Morocco could argue for a solution similar to that
obtained by France in the case of the Channel
Islands if Spain insists on retaining these islands.
(27)

There is a similar problem between Italy and
Malta and the Maltese would like the Italian Pelagic
islands, in the Channel between Malta and Tunisia to
be discounted in drawing the boundary so that
Malta's claim can be extended westward. Italy is not
ready to make the same concessions to Malta that
were granted to Tunisia. (28) This is the reason why
the two countries have only a 'working arrangement'
(Table 13.2).

Second, there are problems of delimitation
which arise because territory controlled by one
country is claimed by another. Gibraltar has been in
British hands since 1704 and Spain wishes to regain
control of it. Spain also retains Ceuta and Melilla

219

in North Africa and if Gibraltar was transferred to
it Spanish territorial waters would cover the
eastern approaches to the Strait. This may explain
why Morocco has threatened to raise the question of
Ceuta and Melilla if Spain continues to press its
claims over Gibraltar. (29) There is no evidence to
suggest that Britain will give Gibraltar back to
Spain soon, or that Spain will agree to negotiate
maritime boundaries between Spanish and British
waters. (30) There is also Cyprus, which has been de
facto partitioned into two parts controlled by Greek
and Turkish-Cypriots and it is impossible that the
Cypriot authorities will agree to draw any maritime
boundaries with Turkey.

The 1979 Peace Treaty signed between Egypt and
Israel has not yet led to any negotiations between
the two states on their maritime boundary. The two
countries are still occupied with the delimitation
of their land boundary and with discussions on
autonomy plans which will determine the future of
the Gaza-Strip including the problems of a maritime
boundary.

RIGHTS OF NATIONAL AND ALIEN VESSELS AND PLANES IN CLAIMED MARITIME ZONES IN THE MEDITERRANEAN

The major rights of alien and national vessels
and planes are: navigation rights, overflight
rights, fishing rights, scientific research, laying
submarine cables, mining, and imposition of
environmental legislation. (31) Alien vessels have
no rights in the internal waters of a state, but
have the right of innocent passage in the
territorial waters of a state and in the contiguous
zone. There are restricted overflight rights in the
territorial waters, and full overflight rights in
the contiguous zone, continental shelf and exclusive
economic zones. Table 13.3 shows some of the common
practices of Mediterranean countries in relation to
alien and national rights.

Contiguous zone. - Only three countries have issued
specific decrees proclaiming their contiguous zones
according to the specifications of the 1982
Convention. Thus, Egypt, Malta and Morocco (in 1981)
expanded their contiguous zones to the permitted
24nm. Syria used an existing decree to extend its
contiguous zone to 6nm more than its territorial sea
limit of 35nm. France, Italy, Spain and Lebanon did
not change old legislation which specified that

Table 13.3: Restrictions in Claimed Maritime Zones in the
Mediterranean

Country	Contiguous Zone	Fishing Zones	Security Zone	Exclusive Economic Zone
Albania	Nil	15nm	special authorisation needed for foreign warships	-
Algeria	Nil	12nm	military vessels must request authorisation 15 days prior to entry	-
Cyprus	Nil	12nm	-	-
Egypt	24nm	12nm	foreign warships require undefined prior notification. Security zone of 18nm (1958)	
France	12nm	12nm	-	200 nm (Atlantic)
Greece	Nil	-	Restricts over-flights of aircraft (10nm - 1931)	-
Israel	Nil	-	-	-
Italy	12nm	12nm	-	-
Lebanon	10.8nm	6nm	-	-
Libya	Nil	12nm	-	-
Malta	24nm	25nm	-	-
Monaco	Nil	12nm	-	-
Morocco	24nm	70nm (Gibraltar Str.- 6nm)	-	200nm
Spain	12nm	12nm	-	200nm (Atlantic)
Syria	41nm	12nm	foreign warships must obtain permission to 41nm	-
Tunisia	Nil	12nm	-	-
Turkey	Nil	-	foreign warships must provide notice prior to transiting	-
Yugoslavia	Nil	12nm	no more than 3 war ships of the same flag may transit at one time (1948)	-

Sources: J.R.V. Prescott, The Maritime Political Boundaries of the World, (Methuen, London, 1985), 299.
R. Smith, (ed.) National Claims to Maritime Jurisdictions, Department of State, Limits in the Seas 36, 5th Rev. (Washington, March 1985).

their contiguous zones reached 12nm. These countries, as signatories to 1982 Convention may adopt extended contiguous zones. All the countries which issued contiguous zones decrees exercise their rights in customs and criminal matters in these zones.

Fishing and Exclusive Economic Zones. The 1982 Law of the Sea Convention made very significant changes in the ability and right of states to use marine resources. The Convention has established the concept of an Exclusive Economic Zone of 200nm in which states have the right to use all the natural resources, certain jurisdiction over scientific research, and environmental protection.

By world standards, Mediterranean fish stocks and catches are small, but its economic value and its relative importance for countries such as Spain, Italy and Greece is high. (32) As Table 13.3 shows, fifteen Mediterranean states have adopted special legislation concerning exclusive fishing zones. Algeria, Cyprus, Egypt, France, Italy, Libya, Monaco, Spain, Syria, Tunisia and Yugoslavia claim fishing zones which correspond to their territorial sea, ie 12nm.(33) Malta has claimed 25nm exclusive fishing zone and Lebanon has a very old (1921) decree relating to fishing zone. Albania had in the past a fishing zone decree which was superseded by a recent decree which expanded its territorial sea to 15nm. (34)

An Exclusive Economic zone of 100nm has been declared by Egypt and Morocco for the Mediterranean, and by Spain, France and Morocco for the Atlantic Ocean. (35) A proclamation of all the littoral states of 200nm will create many overlapping zones and probably will increase the number and frequency of conflicts among the states. The Spanish fishing industry had suffered a marked decline since 1977, when the 200 mile EEZ was first introduced. Spain, Greece and Italy used to fish off North African coasts and the Moroccan expansion of its fishing zone to 70nm reduced their fishing catches and brought Spain and Morocco into serious conflict and even as far as exchanges of fire between naval vessels. Serious disputes on fishing rights took place between Tunisia and Italy in 1975 and 1982 and between Italy and Yugoslavia in 1984. Albania has arrested French and Greek fishermen in its territorial waters. (36) Exclusive economic zones give coastal states the right to control scientific research and to manage the environment, which may eventually prove inducements to their widespread proclamation. (37)

CONCLUSIONS

The application of the 1982 Law of the Sea Convention within the Mediterranean Sea points to possible areas of conflict. The exceptional territorial sea claims of Syria and Albania prevent boundary delimitation with their neighbours and may lead to an overt conflict. Continental shelf delimitation is hampered by potential oil and gas fields in the Northern Aegean Sea and in the area between Libya and Tunisia, and Malta and Italy. The expansion of the exclusive fishing zones in the Mediterranean has produced many fishing disputes which are unlikely to easily disappear. The cultural and economic contrasts of the Mediterranean basin, its history of colonial heritage and political conflicts, makes reaching agreements a very difficult task. Even within the EEC framework, members cannot overcome their overriding differences concerning maritime boundaries though members of the EEC are involved in all the maritime boundary agreements and working arrangements in the Mediterranean.

NOTES

1. N. Kliot, 'The Unity of Semi-Landlocked Seas', in Y. Karmon, A. Shmueli and G. Horowitz, (eds), The Geography of the Mediterranean Basin, (Ministry of Defense, Tel Aviv, 1983), (in Hebrew), 33.

2. G.H. Blake, See Chapter 1 above - See also J.R.V. Prescott, The Maritime Political Boundaries of the World, (Methuen, London; 1985), 295.

3. G.H. Blake, 'Offshore Jurisdiction in the Mediterranean', Paper presented to the Fifth Mediterranean Conference at Bar-Ilan University, Ramat-Gan, Israel, (1980).

4. Nazih N.M. Ayubi, 'The Arab States and Major Sea Issues' in G. Luciani (ed.) The Mediterranean Region - Economic Interdependence and the Future of Society, (Croom Helm, Beckenham, 1984), 127.

5. J.R.V. Prescott, op.cit. (1985) 95-103.

6. J.R.V. Prescott, op.cit. (1985) 61.

7. J.R.V. Prescott, op.cit. (1985), 298. See also R.W. Smith, National Claims to Maritime Jurisdictions, Limits in the Seas No.36, 5th Revision. (U.S. Department of State, Bureau of Intelligence and Research, Washington, 1985), 56.

8. J.R.V. Prescott, op.cit. (1985), 297.

9. J.R.V. Prescott, op.cit. (1985) 298.

10. J.R.V. Prescott, op.cit. (1985) 1.

11. S.P. Jagota, Maritime Boundary, (The Hague, Martinus Nijhoff, 1985), 98.

12. The Geographer, Continental Shelf Boundary: Italy-Tunisia, Limits in the Sea No.89, (Washington DC: Dept of State, 1980), 1-4.

13. The Geographer, Continental Shelf Boundary: Greece-Italy, Limits in the Seas No.96 (Washington DC: Dept of State, 1982), 1-4.

14. The Geographer, Continental Shelf Boundary: Yugoslavia-Italy, Limits in the Seas No.9, (Washington D.C. Dept of State, 170), 7.

15. J.P. Jagota, op.cit. (1985), 116.

16. L.E. Herman (1984) 'The Court Giveth and The Court taketh Away; An Analysis of the Tunisia Libya Continental Shelf Case', International and Comparative Law Quarterly, 33, (4) (1984), 825-858.

17. E.D. Brown, 'The Tunisia-Libya Continental Shelf Case', Marine Policy, (July, 1983), 142-162. M.B. Feldman, 'The Tunisia-Libya Continental Shelf Case: Geographic Justice or Judicial Compromise?', American Journal of International Law, 77(2) (1983) 219-238.

18. Petroleum Economist, January 1985, 30.

19. Petroleum Economist, August 1985, 295; See also Keesing's Contemporary Archives, Vol.30, August 1985, p.33796.

20. J.R.V. Prescott, op.cit. (1985) 87.

21. J.R.V. Prescott, op.cit. (1985) 85.

22. J.R.V. Prescott, op.cit. (1985) 299.

23. G.H. Blake, See Chapter 1 above.

24. A.D. Couper, (ed.) The Times Atlas of the Oceans, (London: Times Books 1983) 186-7.

25. Keesing's Contemporary Archives, Vol.29, April 1984, pp.32791.

26. J.P. Jagota, op.cit. (1985) 164-5.

27. J.R.V. Prescott, op.cit. (1985) 306.

28. J.R.V. Prescott, op.cit. (1985) 306.

29. A.D. Drysdale and G.H. Blake The Middle East and North Africa: A Political Geography (New York: Oxford University Press, 1985) 131.

30. J.R.V. Prescott, op.cit. (1985) 305.

31. J.R.V. Prescott, op.cit. (1985) 40-41.

32. A. Ben-Tuvia, 'Fishing in the Mediterranean' in Y. Karmon, A. Shmueli and G. Horowitz (eds) The Geography of the Mediterranean (Tel-Aviv: Ministry of Defense, 1983) (in Hebrew), 211.
G. H. Blake, 'Mediterranean Non-Energy Resources - Scope for Cooperation and Dangers of Conflict' in G. Luciani (ed.) The Mediterranean Region - Economic Interdependence and the Future of Society (Beckenham, Kent: Croom Helm, 1984) 41-49.

33. J.R.V. Prescott, op.cit. (1985) 299; R.W. Smith, op.cit. (1985), 2-5.

34. R.W. Smith, op.cit. (1985) 12.

35. R.W. Smith, op.cit. (1985) 3-5.

36. G.H. Blake, op.cit. 91984), 41-49. See also Keesing's Contemporary Archives, Vol.30, May 1985, pp.33556-7.

37. G. Luciani, 'The Mediterranean and the Energy Picture' in Luciani, G. (ed.) The Mediterranean Region-Economic Interdependence and the Future of Society, (Beckenham, Kent: Croom Helm, 1984), 5-28.

ACKNOWLEDGEMENT

The author gratefully acknowledges helpful comments and suggestions from Dr Stanley Waterman, and thanks the London School of Economics Cartographic Unit for drawing Figure 13.1.

CHAPTER 14

DEFINING THE INDEFINABLE: ANTARCTIC MARITIME BOUNDARIES

Gerard J. Mangone

PHYSICAL CHARACTERISTICS OF ANTARCTICA

Of the seven great land masses on the planet earth, Antarctica is unique. Although not the largest continent, it is enormous, with an area greater than Brazil, the United States, or Canada. It is twice the size of Europe and equal to all of South America without Argentina. Moreover, it is the coldest, driest, and highest continent. Except for a few coastal areas, all of Antarctica is subject to sub-freezing weather all year round. At the McMurdo Station, which is relatively warm, there are some days in January when the temperature rises above 0 degrees centigrade, but year round the temperature averages -15 degrees and in winter -25 degrees. On average Antarctica has an altitude of about 2,700 meters compared to the average altitude of 1,000 metres for Asia. The altitude makes it about 15 degrees colder on average than the polar region in the north. (1)

Only two or three percent of this vast continent shown on Figure 14.1 is exposed with a rocky surface. All the rest is covered by an ice sheet, dense and heavy, with an average thickness of 2,000 meters. The ice sheet contains about 90 percent of all the world's ice and by comparison is seven times larger than the ice covering Greenland. This vast sheet of ice is nourished by very little precipitation, for the snow that falls on Antarctica is only equal to about 15 cm of water annually. The snow, of course, never melts, and gradually sinks and solidifies into the ice sheet. The ice sheet covering the continent presses down and outward to the shelves at the perimeter of the land mass, inching its way to the sea, often with crevasses 100 meters deep. There is some break-off of ice at the

227

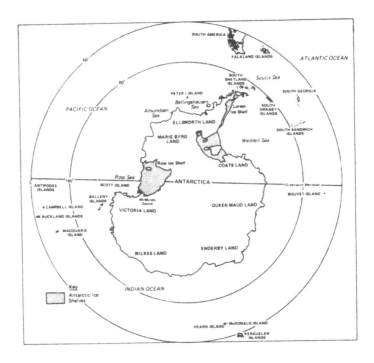

Figure 14.1. Antarctica

sea front and some melting of ice under the sheet, with some snow blown away, so that the total gain in mass from new snow is balanced by the losses.

The continent of Antarctica, far from the centres of world population, utterly forbidding in its climate and terrain, surrounded by stormy seas filled with pack ice or frozen seawater nourished by snow, and by giant icebergs broken from the ice shelf, hardly intrigued the early explorers, except for the hunters of seals or whales in the summer season. But 17 states had research facilities in Antarctica in 1986. During the winter more than 900 people from 13 states have lived and worked at 36 bases in frightful weather, and during the summer the population on the continent has risen to more than 3,000. Moreover, Antarctica has recently become a global political issue, first placed on the agenda of the United Nations General Assembly in 1983, and

its   legal   status   has   become   increasingly
contentious.

DISCOVERIES OF ANTARCTICA

Francis Drake in the sixteenth century had
shown that South America terminated in the southern
ocean; Dutch explorers in the seventeenth century
had visited Australia, Tasmania, and New Zealand
while French navigators had sailed almost to 60
degrees south latitude, discovering some islands.
Yet no one could prove that a land mass existed in
the southern ocean. Captain James Cook, therefore,
was dispatched by the British government to sail as
far south as possible and seek information of value
to commerce and 'natural knowledge'. In 1775 he
discovered the South Georgia and South Sandwich
islands. Although his ship was the first to cross
the Antarctic circle and circumnavigate the
continent, Cook did not discover Antarctica. Forty-
four years later the South Shetland islands were
discovered by Captain William Smith, and in the same
year, 1819, Thaddeus von Bellinghausen was sent by
the Tsar of Russia to the southern ocean. In two
voyages Bellinghausen crossed the Antarctic circle a
few times, and spent almost two months south of 60
degrees south latitude line. He probably sighted but
did not land upon the continent. Claims that Captain
Edward Bransfield, an Englishman, or Captain
Nathaniel Palmer, an American, discovered the mainland in 1820 have been disputed. According to
British sources, the South Orkney islands were
discovered by Captain George Powell in 1821.

Although Captain James Weddell made three
voyages to the Antarctic, reaching 74 degrees south
latitude in the sea that has since been named for
him, and although the voyages of John Briscoe, John
Bellamy, Charles Wilkes, Dumont d'Urville, and
others deserve honors for their approaches to
Antarctica, it was James Clark Ross who clearly
sighted the mainland on the morning of 11 January
1841, with the lofty mountains covered by snow at
Cape Adare. The following morning some twenty-four
men from his vessel landed on a gravelly beach at 77
degrees south latitude with masses of ice and named
it Possession Island. Ross then sailed further south
and discovered the indentation of McMurdo Sound, at
77 degrees south latitude at the edge of the great
ice shelf that bears his name and the volcanic Mt
Erebus. (2)

Antarctic Maritime Boundaries

There was little public interest in Antarctica after the Ross expedition. Another 53 years elapsed before men, in search of whales, actually set foot upon the mainland of Antarctica. In 1895 Johan Bull, Carsten Borchgrevink, Captain Kristensen of Norway and a few crewmen landed on the mainland at Cape Adare, near Possession Island. Four years later Borchgrevink, sailing under a British flag, arrived at Cape Adare again and landed two huts, 75 Siberian dogs, and ten men, who spent the winter at Robertson Bay.

The heroic age of Antarctic exploration then began. It was studded with names like Scott, Amundsen, Shackleton, and Wilson, in the early years of the twentieth century, using ponies and dogs, as they walked across the trackless ice. Later Byrd, Wilkins, Ronne, and Ellsworth, with the advantage of the airplane, could for the first time adequately map the frozen continent. The first good map of Antarctica was printed in Australia in 1939. Better ones appeared in New Zealand in 1953, in Germany in 1954, and in the United States in 1955.

BRITISH CLAIMS TO ANTARCTICA

With the increase of interest in Antarctica at the turn of the twentieth century, Great Britain made the first formal territorial claim in 1908 by declaring that the South Georgia, South Orkney, South Shetland, and Sandwich islands, and the territory known as Graham's land south of the 50th parallel south latitude and between the 20th and 80th degrees of west latitude were dependencies of the Falkland Islands. The letters patent, which encompassed the southern tip of South America, were amended in 1917 to reduce the incursion into Chilean-Argentine areas, but also broadened to include all islands and territories whatsoever south of the 50th parallel between 20 and 50 degrees west longitude and south of the 58th parallel between 50 and 80 degrees west longitude. In 1962 Britain divided the dependencies into (1) the Falkland Island and its dependencies (South Georgia, Sandwich) north of the 60th parallel and (2) the British Antarctic Territory, south of the 60th parallel.

In 1923, moreover, the British created the Ross Dependency as part of the New Zealand dominion, including all islands and territories between 160 degrees east longitude and 150 degrees west longitude south of the 60th degree of south latitude. And in 1938, taking account of the French claim to a slice of Antarctica from 136 to 142 degrees east longitude, the British confirmed Australia's Antarctic Territory as extending from the 45th to the 160th degree of east longitude, south of the 60th degree south latitude, with the exception of the French claim, as shown in Figure 14.2.

The description of the British claims by meridians and longitudes, moreover, led to the extension of these claims by straight lines directly to the South Pole in the form of sectors. This method of making territorial claims, not supported by customary international law and vigorously disputed when advanced in the Arctic region, has been followed by France, Argentina, and Chile, but not by Norway. Although a Norwegian was the first to reach the South Pole in 1911, Norway made no formal claims to Antarctica until 1939 when Germans began staking out claims between the British and Australian sectors. Norway's claim referred only to the mainland coast and the 'environing sea' between the Falkland's Island Dependencies in the west and the Australian Antarctic Dependency in the east.

DISPUTED AND INCHOATE CLAIMS IN ANTARCTICA

Argentina first protested the British claim to Antarctic territories when the 1939 map was published in Australia. Argentina also exchanged a number of notes with Chile over their respective boundaries extending into Antarctica. In 1940 Chile proclaimed all lands, islets, reefs, glaciers and pack ice, already known or to be discovered in the sector between 53 and 90 degrees west longitude to be under Chilean sovereignty. This claim not only overlapped the British claim, but also infringed upon the Argentine claim, which had been asserted by Buenos Aires as the sector within the meridians 25 and 75 degrees west, south of the 60th parallel south latitude. However, in 1943, in a note to Great Britain, Argentina withdrew its claim to a western boundary of 68 degrees, 34 minutes. Then in 1946 Argentina again changed its claim, this time asserting its boundary as 74 degrees west. A year later Argentina agreed with Chile that they would work together to protect their rights (against Great Britain) in the South American Antarctica pending a friendly settlement of their boundaries. (3)

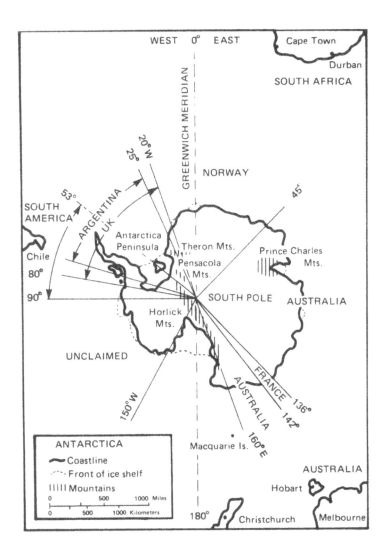

Figure 14.2. Claims to Antarctica

Antarctic Maritime Boundaries

Neither the United States nor the Soviet Union have ever made formal claims in Antarctica. But there can be no doubt about their fundamental interest in the continent and their determination to share in any policy decisions about the future legal structure, economic utility, or environmental protection of Antarctica.

The United States has invested far more of its resources than any other state in the twentieth century in exploration and scientific research in Antarctica. During the 1930s there was evidence that Washington was laying the basis for a territorial claim. (4) After World War II, the United States might have claimed the area from 90 to 150 degrees west longitude, which had not, and still has not, been claimed by any state. However, in the spirit of United Nations collaboration and the concept of trusteeship at that time, Washington approached the claimant states with the idea of internationalising the continent. But nearly all the claimants to Antarctic territory were unreceptive.

For 118 years after the voyages of Belling-hausen to the Antarctic region, neither Tsarist Russia nor the Soviet Union had shown any interest in the continent. Only in 1939 when Norway made its formal claim to part of the Antarctic coast did Moscow react with a note reserving its position about territories discovered by Russia. After World War II, Moscow reacted strongly to the United States initiative with the claimant states to inter-nationalise Antarctica, declaring that any decisions about the future of that continent must take into account not only Russian discoveries, but the Soviet Union's large interests in fishing and whaling there. (5) In brief, the Soviet Union asserted that it must be included in negotiations about Antarctica even though it had not put forth any formal claims to territory.

THE 1959 ANTARCTIC TREATY

On 3 May 1958 President Dwight D. Eisenhower invited 11 states, including the Soviet Union, to discuss an agreement upon a treaty that would give legal effect to the principle of freedom of scien-tific investigation for all persons, organisations, and governments in Antarctica and also ensure that the continent would only be used for peaceful purposes. The President indicated his belief that such a treaty would not require any state to renounce whatever historic rights it might have in Antarctica or to yield any claims to sovereignty it may have asserted. All the eleven states and the United States had demonstrated a remarkable degree of collaboration and cooperation in the quest for scientific knowledge in Antarctica during the International Geophysical Year that had begun in 1957. (6)

The complete demilitarisation of the continent and the guarantee of international scientific cooperation, without any effect upon putative claims to territory and without any administrative structure, appealed to Argentina. Australia, Belgium, Chile, France, Japan, New Zealand, Norway, South Africa, the Soviet Union, and the United Kingdom met in Washington and with the United States signed the Antarctic Treaty on 1 December 1959.

The treaty applies to the area south of 60 degrees south latitude, including all ice shelves. It provides that nothing in the treaty shall be interpreted as a renunciation by the state parties of any previously asserted rights or claims to territorial sovereignty, or any renunciation or diminution of any basis for a claim due to any of their activities or any of their nationals. Nothing in the treaty prejudices the recognition or non-recognition of any other state's right or basis for a claim in Antarctica. Finally, while the treaty is in force, no acts may constitute a basis for asserting, supporting, or denying a claim to territorial sovereignty, and no new claim or enlargement of an existing claim may be asserted.

The treaty may be modified or amended at any time, but it requires a unanimous vote of the contracting parties that are entitled to consultative status: namely, the original twelve contracting parties and by 1986 Poland, West Germany, India, Brazil, China, and Uruguay, which have also been admitted to consultative status. However, thirty years after the treaty has been in force, that is 23 June 1991, any member with consultative status may request a conference of all the contracting parties. Any modification or amendment to the treaty then must be approved by a majority of the contracting states present, including a majority of the contracting states with consultative status. Entry into force, moreover, requires the unanimous consent of the contracting states with consultative status. If contracting states, other than those with consultative status fail to ratify the modification or amendment within two years, they shall be deemed to have withdrawn from the treaty.

In sum, the treaty sought to suspend all affirmations of claims or the modification, enlargement, denial, or projection of claims to territorial sovereignty on the continent, maintaining the indeterminate legal status of the continent by the original contracting states and such other states that were acknowledged by them to have substantial scientific interests in Antarctica, evidenced by investment in research and polar stations. (7)

BOUNDARY PROBLEMS

The only organisational structure created by the 1959 Antarctic Treaty was the meeting of 'the consultative parties'. Since 1961 there have been thirteen regular consultative meetings held in rotation at the capitals of the consultative states. They have largely dealt with scientific research and environmental protection, usually with agreed measures to be taken by their governments.

Even with the best intentions, however, agreed measures, which may involve restrictions such as the prohibition of activities that damage the flora or fauna, run into legal difficulties. Where no territorial sovereign is recognised in the area or in an area that has not even been claimed by any state, punishment for local offences is dubious, not only for the nationals of the state that tries to prosecute, but especially for the nationals of another state who may be charged with violating an agreed measure.

CONSERVATION OF LIVING RESOURCES IN ANTARCTICA

Moreover, as the consultative states turned their attention to conservation of the living resources in Antarctica, especially seals and fish, they were limited in measures to enforce their good intentions without boundaries of national jurisdiction. In 1972 the Convention for the Conservation of Antarctic Seals, not strictly under the Antarctic Treaty system but done by the same states that were members of the consultative meetings, was applied to six species of Antarctic Seals. (8) It provided a permissible catch for three species and forbade the catch of the other three species. Yet there was no allocation of catch by states, no inspection system, and enforcement rested solely in the jurisdiction of the individual states over their own nationals. Most significant, however, was the fact that the members of the consultative meetings had appointed themselves as guardians of the seals in the area south of the 60th degree south latitude although the Antarctic Treaty itself states that nothing in it shall prejudice or affect the rights of any state under international law with regard to the high seas within that area.

In 1980, moreover, having assumed a collective responsibility for the conservation of living resources in Antarctica, the consultative states, with a few additional states invited by them, brought forth the Convention for the Conservation of Antarctic Marine Living Resources. For the first time, however, consultative states extended their agreement to areas slightly north of the 60th degree south latitude. The convention used an ecological boundary, a line of convergence where the warmer, lighter waters flowing south meet with the colder, heavier waters from Antarctica moving north. To avoid a shifting boundary, points ranging from 60 degrees south latitude to 45 degrees south latitude were used for drawing a definitive line. (9) This boundary, however, foreshadowed some of the political difficulties that were bound to arise in Antarctica due to the conflict between recognised boundaries and putative boundaries.

During the course of the Third United Nations Law of the Sea Conference after 1974, states had begun to establish exclusive economic zones extending 200 nautical miles from their shores in which they assumed jurisdiction over all marine resources. Such measures, of course, applied to their overseas territories. British, Norwegian, South African, French, and Australian islands north of the 60th south latitude line were included in the Antarctic Convergence. These well-recognised territories were not affected by the suspension of claims to mainland territories under the Antarctic Treaty. As legitimate territories of sovereign states, the islands may also have an exclusive economic zone. Thus, the French Crozet and Kerguelen islands had an exclusive economic zone, yet these same waters fell under the conservation scheme of the Convention for Antarctic Marine Living Resources. France insisted on the regulation and policing of these exclusive economic zones by itself. (10) Although South Africa had not proclaimed an exclusive economic zone, nothing in international law prevents it from doing so and applying the zone to its Prince Edward Islands, which lie within the conservation zone but outside the reach of the Antarctic Treaty.

Without international recognition of their claims to the mainland of Antarctica, Argentina, Australia, Chile, France, New Zealand, and the United Kingdom have no right to proclaim an exclusive economic zone in the adjacent sea of that area, although they all have done so, more or less, for their national territories, mainland and islands, and there is little doubt about their desire to do so in Antarctica. The commission established under the living resources convention is charged with conservation measures for the entire area within the convergence, but it can only take substantive decisions by consensus of (a) the contracting parties, (b) the acceding parties engaged in research or harvesting activities, and (c) the regional economic integration organisations that may accede to the convention. Moreover, if any member of the commission feels that it cannot accede to a conservation measure, it may declare itself not bound by that measure.

Finally, because of the lack of any national boundaries in the region, enforcement measures to close fishing seasons, limit the quantity of catch, and so forth are left to the flag state of the vessel that may be in violation of the measures. This raises the question of the application of the convention to those states that fail to accept it, insofar as high seas with no national exclusive economic zone still legally exist within the convergence area.

MINERAL RESOURCES REGIME FOR ANTARCTICA

Far more controversial is the effort of the consultative members of the Antarctic Treaty to establish a legal regime for the exploration and exploitation of minerals within the treaty area. Here the paradox of recongised and putative boundaries is more disturbing because fixed installations in the soil, rather than free moving vessels and fish, are involved.

Since 1982 there have been meetings of a special committee of the consultative states in New Zealand, Germany, Japan, Brazil, France, and Australia to discuss a regime for minerals that may be discovered and exploited in Antarctica. A first draft Convention was prepared by the chairman of the committee, Mr Chris Beeby of New Zealand, in 1983, followed by a West German draft, both of which are negotiating documents only. Nevertheless, they have had to face the reality of insubstantial and controversial boundaries that involve political forces.

The committee has moved cautiously, first, to establish a central body that would determine if some areas in Antarctica should be subject to mineral exploitation and, second, the procedures for particular applications to exploit any approved area. The central body comprising all parties to the regime would have the advice of an advisory or scientific committee. (11)

The key to the exploration of the minerals, however, would lie with an executive or regulatory committee that would regulate and monitor mineral activities in one place or area. How shall this important committee be composed? One approach has been to suggest the inclusion of the state or states that have made claims to the area in which mineral development has been proposed together with the state seeking to develop the minerals by itself or on behalf of its nationals. Another approach has been to suggest the inclusion of the applicant state, states that assert claims in the area, and non-claimant states selected from the commission, so that a balance of four claimant and four non-claimant states would result. Another approach would ensure membership on the executive or regulatory committee of the two states that maintain the largest presence in Antarctica, obviously the United States and the Soviet Union.

All these formulae deny paragraph 2 of Article 4 of the Antarctic Treaty, which states that no 'act taken while the treaty is in force shall constitute a basis for supporting a claim.' To admit that a claimant state must be included in establishing the terms, conditions, and fees for mineral development at a site that falls within its claimed area surely supports its claim to that area.

The lack of recognised national boundaries in Antarctica, moreover, has been greatly highlighted by provisions of the 1982 Law of the Sea Convention. Although not yet in force, once 60 states have ratified the convention and one year has elapsed, its articles will have status in international law. Among other things, the convention provides that states may establish exclusive economic zones and notify the Secretary General of their location. States may also propose the limits of their continental shelves, in accordance with the convention, and submit them to a UN Boundary Commission. Yet if no claim in Antarctica is recognised, there can hardly be an exclusive economic zone or continental shelf. (12)

The most innovative institution of the 1982 Law of the Sea Convention is the International Seabed Authority, which is the organ through which states party to the convention shall organise and control activities in 'the Area'. States may establish exclusive economic zones and continental shelves contiguous to their territories, but may not claim or exercise sovereign rights in the Area. Does the Area then include the seabed surrounding Antarctica up to the beaches? If not, by what right can an exclusive economic zone or a continental shelf be established? Certainly there is some legal basis for maintaining that the area include seas adjacent to the unclaimed Antarctic territory unless the consultative states declare and exercise their joint responsibility for that sector of the continent.

Another contradiction may be the baseline for measuring any resource zone. The Antarctic Treaty provisions apply to all areas south of the 60th degree south latitude, including the ice shelves, but the 1982 Law of the Sea Convention in prescribing methods for delimiting baselines does not mention ice shelves.

DEFINING THE INDEFINABLE

The 1982 Law of the Sea Convention was still far from reaching full legal force in 1986. Some of the major powers, moreover, were likely to hold back on ratification. The draft for the minerals regime in Antarctica was far from formal adoption by the states with consultative status, let alone others. Finally, the forbidding environment of Antarctica, coupled with a glut of oil and a depressed hard minerals market in 1986, hardly made mineral exploitation in the region an economic imperative.

Nevertheless, there are major political tides that have moved relentlessly forward to press against the unique legal status and parochial administration of Antarctica during recent years. Having gained acceptance in the 1982 Convention of the concept that the resources of the seabed beyond national jurisdiction are the common heritage of mankind and can only be exploited through an international organisation, many developing states in the United Nations were bold enough to call for international management of Antarctica and an equitable sharing of Antarctica's mineral benefits. A resolution to this effect was approved by the UN General Assembly on 3 December 1985, much to the chagrin of nearly all the states with consultative status under the Antarctic Treaty. Some of their delegates to the United Nations said they would boycott future debates about Antarctica unless the group of developing states, led by Malaysia, abandoned their confrontational tactics. (13)

The boundaries of Antarctica continue to be anomalous. States like Argentina and Chile are not likely to ever abandon their claims; others like New Zealand and Australia have indisputable regional interests in the continent; and neither the United States nor the Soviet Union, so long as they are superpowers, are inclined to let the future of this vast area of great scientific interest and potential wealth fall into the maelstrom of UN politics.

The challenge will be for the responsible states in Antarctica to assuage the concerns of the developing states that policy decisions will not be openly debated and fully explained to the world community; furthermore, to allay their fears that the continent may be exploited for the benefit of only a few states, which by wealth and chance, obtained favoured positions on a frozen continent over which they could not rightly maintain territorial claims under customary international law.

Much of the future of Antarctica will depend upon the recognition of boundaries, not as exclusive preserves of sovereigns, but as areas of fixed responsibilities. The problem of reconciling national jurisdiction over specific areas with open access for both scientific research and economic exploitation is not insoluble. Antarctica and its surrounding seas, because of their location and extraordinary physical environment, can be treated differently than other areas of the world, allowing some states to be guardians of the world heritage but not denying any state a share in its benefits.

NOTES

1. For the physical characteristics of the continent of Antarctica, see the sixteen-page article in the New Encyclopedia Britannica, Vol.I, 15th edition, (1983), which has a bibliography

divided into Geology, Climate and Meteorology, Glaciology and Oceanography, Biology, and History. See also the Great Soviet Encyclopedia, Vol.2, 3rd edition, English translation, (Macmillan Educational Corporation, 1973) for other details on the physical attributes of the continent. The most comprehensive bibliography on Antarctica is the US Library of Congress, Antarctic Bibliography, Washington DC, published in 14 volumes since 1962 and covering publications from 1951 onward.

2. There are innumerable first-hand memoirs of the participants in the various polar discoveries over the years, such as R. M'Cormick, Voyages of Discovery, 2 vols, (Sampson Law, London, 1884), which includes an account of the Ross expedition, the landing on Possession Island, and the running-up of the British flag. Surveys of the wealth of material on Antarctic discoveries, some contra-dictory, may be found in L.P. Kirman, The White Road, A Survey of Polar Explorations, (Hollis and Carter, London, 1959), and David Mountfield, History of Polar Exploration, (Hamlyn, London, 1974). Another excellent study of the early exploration and exploitation of the Antarctic region through the quest for seals can be found in Briton Cooper Busch, The War Against the Seals, (McGill-Queens University Press, Montreal and Kingston, 1985).

3. A well-organised review of each of the claims to Antarctica was made in John Hannessian, Jr, 'National Interests in Antarctica' in Antarctica, ed. by Trevor Hatherton, London, (1965). See also J.R.V. Prescott, 'Boundaries in Antarctica' in Australia's Antarctic Policy Options, (Centre for Resource and Environmental Studies, Australian National University, Canberra, 1984) for specifics and commentary on the role of ice shelves in boundaries.

4. See Marjorie Whiteman, Digest of International Law, Vol.2, Washington DC, p.1248, for the United States approach to making a claim in Antarctica.

5. See Whiteman, pp.1254-1255 for the Soviet reactions to the American attempt to inter-nationalise Antarctica without consideration of Russian interests.

6. For the text of the 1959 Antarctic Treaty, US press releases, and statements about the treaty by

the President and Secretary of State, see US Department of State <u>Bulletin</u>, 41 (Oct.-Dec. 1959), Washington DC, p.910.

7. The books, some of them multi-authored, plus the articles dealing with the Antarctic Treaty and the several boundary-resource problems arising from that treaty are far too numerous and diverse to list. Among many good works, W.M. Bush, <u>Antarctica and International Law</u>, (Oceana, London, 1982); F.M. Auburn, <u>Antarctic Law and Politics</u>, (Indiana University Press, Bloomington, 1982), and P.M. Quigg, <u>A Pole Apart: The Emerging Issue of Antarctica</u>, (McGraw Hill, New York, 1983), deserve special mention.

8. The text of the Convention on the Conservation of Antarctic Seals is in US Department of State, <u>United states Treaties and Other International Agreements</u> (<u>TIAS</u>), 29, Part 1, Number 8826 (1966-67).

9. The Text of the Convention on Antarctic Living Marine Resources may be found in American Society of International Law, <u>International Legal Materials</u>, 19 (1980), 841. See also, Ronald F. Frank, 'The Convention on the Conservation of Antarctic Marine Living Resources', <u>Ocean Development and International Law</u>, <u>13</u> (1983), 291-346.

10. A statement in the Final Act of the Canberra Conference dealing with the convention notes that France will continue to enforce the regulations in its Exclusive Economic Zone established prior to the effects of the convention, that France may promulgate regulations more stringent than the international commission, and that France alone would police regulations in the zone. These conditions may be applicable by other states with recognised island possessions within the convergence boundary, but excluded from the Antarctic Treaty area.

11. General principles about minerals exploitation in the Antarctic were first set forth by a special preparatory meeting of the consultative parties in Paris in 1976. The next four years saw little progress toward the articulation of a mineral resources regime, but the Eleventh Meeting of the Consultative Parties in Buenos Aires in June-July 1981 brought forth ten recommendations (to their

respective governments) about a mineral regime for the Antarctic. In 1983 Chris Beeby, chairman of the informal meeting of the consultative parties to discuss a minerals regime, drafted a paper of 36 articles for a regime. For the text and commentary by Greenpeace International, which believed that a minerals regime required the participation of states other than the consultative parties, see 'The Future of the Antarctic', Temple House, United Kingdom, 1 October 1983. For a solid study on Antarctic mineral resources, see W.E. Westermeyer, The Politics of Marine Resource Development in Antarctica, (Westview Press, Boulder, Colorado, 1984) and for another analysis, see S.A. Zorn, 'Antarctic Minerals: A Common Heritage Approach, Resources Policy, 10 (1984), 2-18. I am also indebted to a telephone interview with Tucker Scully, US Department of State, the American representative on the Special Committee for Minerals, 15 May 1986.

12. On this point, see Roland Rich, 'A Minerals Regime for Antarctica', International and Comparative Law Quarterly, 31 (1982), 717-18. A thorough discussion of the legal issues involved in the mining of Antarctic mineral resources, with an analysis of the drafts presented by Chris Beeby of New Zealand and by the Federal Republic of Germany can be found in a paper by Francesco Francioni, 'Legal Aspects of Mineral Resource Exploitation in Antarctica', presented to Regional Meeting of the American Society of International Law, Cornell University, 7-8 October 1985.

13. The presentation and discussion of the Antarctic issue in the United Nations has been reviewed extremely well, with documentary references, by Moritaka Hayashi, 'The Antarctic Question in the United Nations' a paper presented to the Regional Meeting of the American Society of International Law at Cornell University, 7-8 October 1985.

CHAPTER 15

BEYOND THE BOUNDS? A CONSIDERATION OF LOCAL
GOVERNMENT LIMITS IN THE COASTAL ZONE OF ENGLAND AND
WALES

Joyce E. Halliday

INTRODUCTION

Local government, at both the County and the
District level, is an important planning and
managerial agency on land. It follows therefore that
it is an important planning and managerial agency in
coastlands where, in England and Wales, some 30
Counties and over 110 Districts adjoin coastal
waters (1) and where proximity to the sea has
prompted numerous area specific responses. The
boundaries of local government exert, moreover, a
pervasive organising influence upon other agencies
active in this coastal area. The jurisdiction of
local government tends, however, to stop at the low
water mark (2) and its place in an exposition on
'maritime' boundaries and 'ocean' resources would
seem therefore, at first consideration, to be
peripheral. This paper aims to challenge this
contention.
    It is suggested firstly that many of the
activities occurring offshore, (particularly in the
nearshore zone), have ramifications for the use and
management of coastal lands, that is ramifications
for the present statutorily enacted concerns of
local government. The converse is, to a degree, also
true. It is myopic therefore to seek progress in the
allocation and control of ocean resources without
considering their relationship to existing systems
of resource management. This is particularly true in
the strategic areas where the systems meet and
merge, in this instance in the coastal zones. The
evolution of maritime boundaries has, therefore,
repercussions onshore.
    A second and related aspect is that the
maritime focus for boundary demarcation is largely
coincident with the oceans beyond the territorial
seas. There remains therefore the still maritime

244

environment and resources of the territorial seas
and internal waters, regions where individual
nations remain the determinants of policy. In
England and Wales this area is largely a managerial
vacuum. It is affected however, as noted above, by
decisions taken in the maritime realm beyond. It is
also affected by a counter force, the extension and
adjudication of primarily land based systems of
control into the maritime margins.

This paper focuses on one such system. It
examines some of the maritime boundaries spawned by
local government, including the forces behind their
conception and their relationship to ocean
resources. It illustrates the fact that local
government jurisdiction possesses a maritime
dimension, or rather several separate maritime di-
mensions. Finally it considers the possible
evolution of this sphere of local government
organisation.

EXISTING LIMITS

Two main organising principles are used. The
first is the nature of the bounded resource. Ocean
resources are defined to include not only the more
commonly discussed factors such as fisheries, oil
and gas, aggregates and minerals, but also other
resources whose control has contributed to local
government intervention. This includes for instance
the quality of coastal waters, their flora and fauna
and the availability of water space for recreation
and communication.

The second is the nature of the boundary
itself, whether it is for instance statutory,
operational or perceptual and how it relates to the
controlled resource. A coincidence of the limits of
law, action, viewpoint and resource, whilst the
strongest type of boundary, is rare in reality.

PORT/HARBOUR AND RELATED LIMITS

The area beyond low watermark where local
government boundaries are perhaps most prolific is
in the field of port and harbour administration.
There are over 40 local authority ports in England
and Wales, with at least 36 District Councils acting
as port authority. The majority are in Wales and the
South West and are managed by a single District
Council.

Their area of jurisdiction, laid down by local act, is not necessarily confined to internal waters but may extend for various distances into the territorial sea. The limits of the port of Sunderland for instance lie 480 metres offshore at Souter Point, and 1095 metres offshore at Ryhope Dene.

Many of these limits are historical in origin, predate local authority involvement and are not related directly to present requirements. Indeed many of these ports now handle only fishing and pleasure craft, activities whose reconciliation can, nevertheless, pose profound managerial challenges. Others such as Sunderland (2,522,435 tonnes in 1983), Colchester (1,070,156 tonnes) and Ramsgate are still commercially important. Limits are again, however, frequently related to the necessity for control in a different era. In Sunderland, for instance, the local authority only became port authority in 1972 with the passing of the Sunderland Corporation Act. Certain of the current boundaries were enacted, however, as early as 1759. The port's jurisdiction extends therefore 9½ miles inland up the River Wear, a residual maritime boundary from another era.

Less absolute but still an important extra-territorial responsibility, is the contribution of local authorities, both Counties and Districts, to variously defined port trusts and conservancies, where they frequently represent a major influence. Eleven of the 15 commissioners on the Chichester Harbour Conservancy, for instance, are elected by local government, as are 11 of the 21 commissioners for Cowes harbour, 6 of the 14 at Lymington and 6 of the 19 at Great Yarmouth. Local authorities are also represented on such major port trusts as the Tyne, Tees and Hartlepool and Port of London Authority. (3)

Boundaries are similarly variable in their offshore extent although, as for instance in Chichester and Langstone Harbours, often limited. Here the offshore dimension, the harbour waters, are effectively bounded by land except for the entrance channels. The bounded administration provides an example, also, of another factor which is a potentially strategic feature in the maritime extension of local authority interests, that is joint administration. In instances this is a joint effort between two or more local authorities, the port of Sunderland for example has until recently been responsible to a committee consisting of members from Sunderland Metropolitan Borough Council and Tyne and Wear County Council. In other areas it involves the local authorities in consort with affected interests, including users. The commissioners for the harbour of Rye for instance include three nominees of the Royal Yachting Association, four to six representatives of commercial interests in the harbour, together with a representative of the Nature Conservancy Council, the Rye Fishermen's Society, the Sea Fisheries District, the County Council and two members of the District Council. (4) There is a recognition therefore that this resource, water space for the movement, loading and mooring of commercial, recreational and fishing craft, needs a correspondingly multi-dimensional response.

A related maritime function with a separate set of boundaries is responsibility for port health. Some 88 District Councils either act as port health authority, or have representatives on the local port health authority. Boundaries are once again individual, being laid down in the constituting orders, but may comprise either the whole or part of a Customs port. Where they are coincident with Customs limits they therefore extend beyond harbour limits discussed above, to the outer boundaries of the territorial sea. Again joint action is frequently a feature of this offshore involvement. Swansea Port Health Authority for instance involves representatives from five district Councils.

Involvement with pilotage is more limited. Only two authorities, Wisbech and Bristol, act as pilotage authority, although there are also examples of local authority representation on the pilotage authority. In the case of Bristol the boundaries are even more extensive than those discussed above, extending more than ten miles beyond territorial limits in the Bristol Channel. (5)

Figure 15.1 illustrates those authorities known to have responsibilities for port or harbour functions involving a maritime extension of control. At a minimum this involves 109 districts, a far from insignificant intervention beyond the bounds. It is important to emphasise, however, that these boundaries represent a series of limits relating to single functions. They are frequently drawn around point resources, that is single ports and because extensions have been drawn in response to one requirement alone, the remaining powers of local government have not been correspondingly extended.

Langstone Harbour, together with neighbouring Chichester, represents a good example for instance of relatively recently enacted harbour limits, drawn

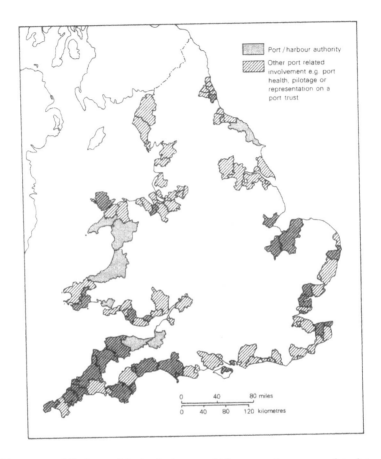

Figure 15.1. Districts with port or harbour functions involving a maritime extension of control

in response to contemporary pressures, predominantly the reconciliation of conservation and recreation on and offshore. (6) One of the arguments used forcibly, however, at a public inquiry into the Langstone Harbour Revision Order, was that the two local authorities forming the Harbour Board had enjoined to create a separate harbour authority, not to extend their jurisdiction as local government agents. The powers exercisable over the waters were therefore claimed to be determined by the enabling legislation alone. It has nevertheless been argued that planning authorities, despite their present reluctance, are entitled to control developments within inlets and estuaries intra fauces terrae (7) and hence within strategic sections at least of many harbours. The exact relationship of local government powers to offshore limits, particularly to harbour bounds, is in many instances therefore blurred. (8)

Perhaps the best example, however, of harbour related action offshore, in accordance with the realities of the contemporary resource and its use, is the activities of the Solent Sailing Conference. Convened by Hampshire Council and now supported by the Isle of Wight, this lacks precise boundaries and has no statutory basis. It extends beyond the concerns of individual sectors or harbours to provide a forum for organisations and individuals with a common interest in the use of the Solent. Here therefore local government has gone beyond the bounds, not to create a new set of unilateral limits but to act as a broker, bringing interests together and breaking barriers down, particularly the barriers of contrasting viewpoint.

Recreation

Recreation, as noted above, has become an important component in the management of many of these harbours, a prime motive in some more recent harbour enactments and a force behind non-statutory initiatives. Some degree of influence over the recreational user of the offshore zone is also gained by the control of landward facilities relating to coastal waters, access points and parking facilities for example. Havant Borough Council for instance have acted to control water skiing by imposing an excess charge on cars towing high speed boats. Ogwr Borough Council (South Wales) feel they have a similar ability to influence behaviour in the nearshore zone without intervening directly. The council own one of only two slipways in the district, the other is club controlled, any infringement by a power boat in coastal waters means therefore that it is subsequently unable to launch in the area. Council controlled moorings have a similar influence where the facility is in short supply.

More specifically local authorities also have the ability to make byelaws for the control of bathing and boating extending 1000 metres to sea.

These powers have not been universally used but many
authorities do have byelaws relating to the coast.
Water safety, including power boating, surfing,
sailboarding, bathing and water skiing, is now the
dominant element amongst such byelaws. Control of
activities in the offshore zone is therefore
increasingly being perceived as an area where local
government needs to take action, particularly where
access to the seas is difficult to control.

Fisheries limits

The mechanism by which local government
interest in the control of fishery resources is
mediate is the Sea Fisheries Committee, the enabling
legislation for which was passed in 1888 with the
aim of securing the proper management of fisheries
in England and Wales. These Committees are allied to
the committee structure of a County Council, as for
instance in Devon, Cornwall and Cumbria, or, more
commonly, form a joint committee of several County
Councils. Membership is drawn not only from the
constituent County Councils but also from the
Regional Water Authorities and from local fishery
interests as appointed by MAFF and the Welsh Office.
Local government action offshore in relation to
fisheries is markedly different therefore from the
harbour and recreational scenarios sketched above.
Firstly it is the province of County Councils.
Secondly it operates on a regional rather than a
point basis, control extending from the whole of the
English and Welsh coastline. Jurisdiction is
therefore comprehensive within, but restricted
still, to territorial waters and thus falls short of
fishery limits. In other areas too the boundaries
fail to coincide with the resource and its
exploitation. The boundary between the Sussex and
the Southern Sea Fisheries District for instance
lies on Hayling Island, dividing the interconnected
harbours of Langstone and Chichester, whilst
problems are always found when a static areal system
seeks to control a mobile, common property resource.
This set of statutory boundaries is interesting
because it is an example of local government having
a controlling influence which extends again beyond
low water mark, this time to the edge of the
territorial sea. It is not an area, experience
suggests, where local authorities feel themselves
particularly involved in coastal zone management but
provides nevertheless a pointer to the kind of
action they could take offshore if other factors,

political will for instance, were present. It is for instance an example of local government acting on a regional basis, exemplified by the Lancashire and Western Sea Fisheries District which extends across five Counties from Cardigan to Morecambe Bay. It is also an example of a local government committee including, by statute, representatives from the other parties with an interest in the resource.

Aggregates

The example of aggregates, in quantitative terms 'the most important hard marine mineral deposit mined', (9) can be used to illustrate a very different point. It is a resource of the whole coastal zone, land, sea and shore, with about 18 per cent of our total national annual production coming from the seabed and the Continental Shelf. (10) Extraction on land and shore is controlled essentially by planning permission and related conditions, issued at County level. Extraction at sea is controlled, in contrast, by the issue of a licence by the Crown Estate Commissioners.

This is an example therefore of an area where local government boundaries have not extended beyond the low water mark and where separate loci of control pose problems for the long term planning and management of the resource. This is particularly pertinent for the reconciliation of effort given that the Crown Estate Commissioners have no policies at present for initiating exploration, but rather welcome proposals. Both the lack of strategy, and their financial involvement, contrast with the guidance given by the planning system on land.

Hydrocarbons

A similar dichotomy is also found with respect to oil where the County Councils are again responsible for determining an application on land whilst offshore it is the province of the Department of Energy. The situation is similar also with gas, whilst offshore pipelines are the responsibility of the pipeline inspectorate of the Department.

The application by the British Gas Corporation to Barrow Borough Council for a reception and treatment terminal for natural gas and the construction and operation of the associated pipeline, provides just one illustration of the consequent inability of local government to look beyond the bounds in this area. (11) Here strategic factors such as the potential impact of the gasfield on fisheries and the maritime impact of the incoming pipeline and seabed operations were acknowledged but could not be taken into account in determining the applications, because they lay beyond the authority's judicial limits.

The coastal configuration of the borough had, however, interesting repercussions, for the pipeline came ashore firstly on Walney Island, parallel to the mainland. It had therefore to cross the intervening Walney Channel before reaching the terminal stie, 3 kilometres inland from the first landfall. This section of the route, including the Walney Channel thus fell within the Council's decision making environment and, in respect of this maritime area between the mainland and the island, the authority indeed considered the whole range of relevant impacts. This is one instance then where a precedent exists for local government to take decisions which attempt to weight the complex of coastal zone activities, including the maritime environment, rather than only those which impinge directly on land use considerations.

The main response in the offshore dimension to hydrocarbons in England and Wales, as opposed to Scotland, has however, been in the field of pollution control. Subsequent to the Torrey Canyon incident local authorities were requested, by circular, to assume responsibility for clean up operations within one mile of the shore and onshore. This function is variously divided between district and county administrations and has been comprehensively acknowledged with the preparation of numerous contingency plans, documents which again provide for the united action of many different agencies in an offshore area.

The Marine Pollution Control Unit, established in 1979 and dedicated entirely to contingency planning and operations, has now also assumed responsibility for the overall supervision and clean up of oil spills around the coastline. This represents therefore a further type of local government maritime boundary, a standard, arbitrary width from the coast, without statutory backing, yet widely upheld. It may also represent another potential trend towards more comprehensive coastal zone management, a centralised system of strategic control superimposed upon a divided and essentially local responsibility.

CONCLUSION

This paper has shown that local government control although commonly held to end at the low water mark has a maritime dimension. This dimension is multi-faceted; it varies, depending on purpose and locality, in width, effective occupance, uptake and status.

In addition to such offshore limits local government also maintains a degree of control over maritime activities, particularly those in the near-shore zone, via the provision or otherwise of on-shore facilities. Conversely however, it is increasingly impinged upon by actions occurring offshore, which lie beyond its control yet are frequently intimately associated with a corres-ponding set of actions on land.

The maritime dimension is also therefore incom-plete. The enforced partiality of the consideration of, and the solutions proposed to, the problems of colliery waste tipping on the Durham coastline, (12) is evidence again of, inter alia, the persistence of the low water mark cut off point for many areas of local government activity. Attention is thus focused on the shore although, in an interconnected system, waste now extends far beyond, out to the 10 fathom line and to a depth of 0.8 metres below the seabed surface. There are many such areas where a maritime dimension, not necessarily bounded by statute but at least at the level of a co-ordinating forum, is now required.

There is evidence that the perception of the offshore zone, by local authorities, is also changing. In the Isle of Wight for instance the County is concerned about the increasing problem of sea-borne hazardous substances arriving on its shores and the oil pollution risk. The Council have also established a standing conference, including the Borough Councils, the water authority, other statutory undertakers, Trinity House, the Department of the Environment and the Crown Estate Commissioners, to examine the problems associated with the island's coastline. This is addressed particularly at the inter-relationship between rates of erosion and the marine dredging for aggregates which takes place at varying distances offshore. In 1983 the Council was made aware of yet another potential marine impact when there were proposals to moor a liquified petroleum gas tanker in the Solent, off the Isle of Wight.

Similar examples are common around the coast,

not manifest necessarily in offshore boundaries, but in an increasingly permeable perceptual boundary, which looks beyond the coastline and seeks formally, or more frequently informally, to exercise a degree of control over the adjacent maritime environment.

The administrative realm of local government at the coast can therefore be summarised as a jurisdiction which ceases generally at low water mark, a series of offshore boundaries related to the control of individual resources, a reciprocal influence imposed by mere adjacency and an increasing awareness of the maritime environment beyond.

It has been suggested that current activity by local government at the maritime margins should now be consolidated by the extension of planning control three miles out to sea, (13) a distance which was presumably seen as a proxy for territorial limits. Such a development remains always a possibility, not least because the management of the territorial sea remains more akin to land, 'to a large extent concerned with the rational exploitation of an area under the exclusive domestic jurisdiction of one state', (14) than to the oceans beyond. Problems, however, are manifold, including the fact that, because of bay closing lines, the distances offshore would, in many areas, be considerably greater than three miles. It would therefore require a fundamental leap by the land use planner, who would also have to adjust to a three-dimensional reality, adjudicating between uses of the seabed, waters and their surface. It would also potentially result in the balkanisation of coastal waters and create a set of authorities whose responsibilities would be fundamentally different from those of their inland counterparts, with profound implications for local authority finance and accountability.

Extrapolation from the present base would suggest instead only incremental change in these critical margins, with continued creation and consolidation of offshore boundaries in strategic areas such as major estuaries and harbours. Here the need for change is frequently both apparent and acknowledged. An imposition of new local government frontiers without this corresponding political will to volunteer as maritime frontiersmen, is unlikely to produce a successful venture beyond the bounds, particularly in an era of constrained resources. The territorial sea meanwhile, an attractive 'visible' boundary, bears little relationship to the current view of the maritime realm as seen from council offices. Concern for the offshore zone, whilst expanding, remains much closer to our shores.

NOTES

1. The author defined 128 'coastal zone' districts and 30 'coastal zone' Counties in an unpublished survey of local authority action and attitude at the coast. As a minimum for instance, 110 districts have coast protection functions. The same survey is source for many other comments on local authority involvement, unless otherwise attributed.

2. Low water mark in England and Wales being the mean low water mark of ordinary tides. See for instance J. Gibson, 'Foreshore: A concept built on sand', Journal of Planning and Environmental Law, (1977), 762-770.

3. See, for example, G. Adams, Organisation of the British Port Transport Industry, (London: National Ports Council, 1973) and R.P.A. Douglas, Harbour Law, (Lloyds of London Press Ltd, 1983, 2nd Edition).

4. Harbour of Rye Revision Order, 1976.

5. J. Gibson, 'Coastal Zone Management Law: A case study of the Severn Estuary and the Bristol Channel', Journal of Planning and Environmental Law, (1980), 153-165.

6. See The Chichester Harbour Conservancy Act of 1971 and the Langstone Harbour Act of 1962.

7. J. Gibson, 'Coastal Zone Management Law', Paper presented to the Greenwich Forum Conference 1962, 'Perils and Prospects of the present inshore and offshore regime for Britain's coasts and coastal zone. Is it time for a change?', Greenwich VIII, London, 1962.

8. In the case for instance of Harwich Harbour Conservancy Board v. The Secretary of State for the Environment (1975) the Court of Appeal ruled that the Harbour Authority did not have the powers to refuse an application for additional moorings on the grounds that the harbour had insufficient navigational capacity; only the planning authority (ie local government) could exercise such control.

9. A.D. Couper (ed.), The Times Atlas of the Oceans, (London: Times Books, 1983).

10. The Crown Estate Commissioners, The Crown Estate: Report of the Commissioners for the year ended 31.3.83, (London: HMSO, 1983).

11. See, for example, Borough of Barrow in Furness, A report to the planning committee on an application for planning permission for construction and operation of a 914mm diameter pipeline between low water mark, west of low bank, Walney Island (crossing Walney Island, Roosecote Sands and Walney Channel) to Barrow onshore terminal, Westfield Point, Barrow-in-Furness. (Unpublished, 1981).

12. The Durham Coast, (Durham County Council/ Easington District Council, 1982).

13. D.F. Shaw, 'Conservation and Development of Marine and Coastal Resources', In The Conservation and Development Programme for the UK, (London: Kogan Page, 1983).

14. E.D. Brown, 'Sea Use Planning in the North Sea: The Legal Framework', in J.K. Gamble (ed.) The Law of the Sea: Neglected Issues , in Proceedings of the 12th Annual Conference of the Law of the Sea Institute, the Hague, 1978, (Law of the Sea Institute, Hawaii, 1979).

SELECT BIBLIOGRAPHY

The literature on maritime boundaries and ocean resources is large and expanding rapidly. This bibliography is selected from the sources cited, with some additional references.

Adams, G. Organisation of the British Port Transport Industry. (London: National Ports Council, 1973).

Alexander, L.M. 'Offshore claims and fisheries in North-West Europe' Yearbook of World Affairs XIV (1960) 236-260.

Alexander, L.M. Offshore Geography of Northwestern Europe: The Political and Economic Problems of Delimitation and Control. (Chicago: Rand McNally and Co., 1963; London: John Murray, 1966).

Alexander, L.M. 'Baseline delimitations and maritime boundaries' Virginia Journal of International Law 23 (1983) 503-536.

Alexander, L.M. 'The delimitation of maritime boundaries' Political Geography Quarterly 5 (1986) 19-24.

Antarctic Bibliography. 14 Vols. (Washington, DC: Library of Congress, 1962-1986).

Arbitration Between the United Kingdom of Great Britain and Northern Ireland and the French Republic on the Delimitation of the Continental Shelf. Decisions of the Court of Arbitration 30 June 1977 and 14 March 1978 Misc. No.15, Cmnd.7438. (London: HMSO, 1978) reprinted as 'France-UK: arbitration on the delimitation of the continental shelf' in International Legal Materials XVIII (1978) 397-494.

Select Bibliography

Archer, A.A. and P.B. Beazley 'The geographical implications of the Law of the Sea Conference' Geographical Journal 141 (1975) 1-13.

Archer, C. and D. Scrivener 'Frozen frontiers and resource wrangles: conflict and co-operation in northern waters' International Affairs 59 (1983) 59-76.

Auburn, F.M. Antarctic Law and Politics (Bloomington: Indiana University Press, 1982).

Australia's Antarctic Policy Options (Canberra: Centre for Resource and Environmental Studies, Australian National University, 1984).

Ayubi, N.N.M. 'The Arab states and major sea issues' in G. Luciani (Ed.) The Mediterranean Region: Economic Interdependence and the Future of Society 126-148. (London: Croom Helm, 1984).

Barston R.P. and P. Birnie (Eds) The Maritime Dimension. (London: George Allen and Unwin, 1980).

Beazley, P.B. Maritime Limits and Baselines: A Guide to Their Delineation. (Second Edition) (The Hydrographic Society, Special Publication No.2, 1978).

Beazley, P.B. 'Half-effect applied to equidistance lines' International Hydrographic Review 56 (1979) 153-161.

Beazley, P.B. 'Maritime boundaries' International Hydrographic Review 59 (1982) 149-159.

Ben-Tuvia, A. 'Fishing in the Mediterranean' in Y. Karmon, A. Shmueli and G. Horowitz (Eds) The Geography of the Mediterranean Basin, (Tel Aviv: Ministry of Defence, 1983), (in Hebrew).

Bird, E.C.F. and M.L. Schwartz The World's Coastline, (New York: Van Nostrand Reinholt Company, 1985).

Blake, G.H. 'Offshore jurisdiction in the Mediterranean' Ekistics 48 (1981) 339-344.

Blake, G.H. 'Mediterranean non-energy resources: scope for cooperation and dangers of conflict' in G.

258

Select Bibliography

Luciani (Ed.) The Mediterranean Region: Economic Interdependence and the Future of Society 41-74. (London: Croom Helm, 1984).

Blecher, M.D. 'Equitable delimitation of the continental shelf' American Journal of International Law 73 (1979) 60-88.

Boggs, S.W. 'Delimitation of seaward areas under national jurisdiction' American Journal of International Law 45 (1951) 240-266.

Booth, K. Law, Force and Diplomacy at Sea (London: George Allen and Unwin, 1985).

Borgese, E.M. and N. Ginsburg (Eds) Ocean Yearbook(s) 1-5. (Chicago: Chicago University Press, 1981-1985).

Bowett, D.W. The Legal Regime of Islands in International Law. (Alphen aan den Rijn, The Netherlands: Sijthoff and Noordhoff. Dobbs Ferry, New York: Oceana Publications, Inc., 1978).

Bowett, D.W. 'The Arbitration between the UK and France concerning the continental shelf boundary in the English Channel and South Western approaches' British Yearbook of International Law XLIX (1978) 1-29.

Bozcek, B.A. 'Global and regional approaches to the protection and preservation of the marine environment' Case Western Reserve Journal of International Law 16 (1984) 39-70.

Broadus, J.M. and P. Hoagland III 'Rivalry and coordination in marine hard minerals regulation' in Exclusive Economic Zone Papers 55-61, (Rockville, Maryland: National Oceanic and Atmospheric Administration, 1984).

Brown, E.D. The Legal Regime of Hydrospace (London: Stevens, 1971).

Brown, E.D. 'The continental shelf and the exclusive economic zone: the problems of delimitation at UNCLOS III' Maritime Policy and Management 4 (1977) 377-408.

Brown, E.D. 'Rockall and the limits of national jurisdiction of the UK', Marine Policy 2 (1978) 181-

Select Bibliography

211.

Brown, E.D. 'Rockall and the limits of national jurisdiction of the UK, Part 2' Marine Policy 2 (1978) 275-303.

Brown, E.D. 'Sea use planning in the North Sea: The legal framework' in J.K. Gamble, Jnr (Ed.) The Law of the Sea: Neglected Issues, (The Hague, 1979).

Brown, E.D. 'The Anglo-French continental shelf case' San Diego Law Review 6 (1979) 461-530.

Brown, E.D. 'The Tunisian-Libyan continental shelf case: a missed opportunity' Marine Policy 7 (1983) 142-162.

Brown, E.D. 'The UN convention on the law of the sea 1982' Journal of Energy and Natural Resources Law 2 (1984) 258-282.

Brown, E.D. Sea-bed Energy and Mineral Resources and the Law of the Sea: Volume 1; The Areas within National Jurisdiction. (London: Graham and Trotman, 1984).

Buchanan, N. and D. Steel 'Meaningful effort limitation: the British case' in Fisheries of the European Community. (Edinburgh: White Fish Authority, 1977).

Burmester, H. 'The Torres Strait Treaty: ocean delimitation by agreement' American Journal of International Law 76 (1982) 321-349.

Busch, B.C. The War Against the Seals. (Montreal and Kingston: McGill Queens University Press, 1985).

Bush, W.M. Antarctica and International Law. (London: Oceana, 1982).

'Case concerning delimitation of the maritime boundary in the Gulf of Maine area (Canada v United States), Judgement of October 12, 1984) Reports of Judgments Advisory Opinions and Orders, 1984 246-390. (The Hague: International Court of Justice, 1984) Reprinted in International Legal Materials XXIII (1984) 1197-1273.

'Case concerning the continental shelf (Tunisia/

Select Bibliography

Libyan Arab Jamahiriya), Judgment of 24 February, 1982' Reports of Judgments Advisory Opinions and Orders, 1982 18-323. (The Hague: International Court of Justice, 1982) Reprinted in International Legal Materials XXI (1982) 225-317.

'Case concerning the continental shelf (Libyan Arab Jamahiriya/Malta), judgment of 3 June 1985' Reports of Judgments, Advisory Opinions and Orders, 1985. (The Hague: International Court of Justice, 1985). Reprinted in International Legal Materials XIV (1985) 1189-1276.

Charney, J.I. 'Ocean boundaries between nations: a theory for progress' American Journal of International Law 78 (1984) 582-606.

Chircop, A.E. and I.T. Gault 'The making of an offshore boundary: The Gulf of Maine case, 1984' Oil and Gas Law and Taxation Review 7 (1984/85) 173-181.

Churchill, R.R. 'Maritime delimitation in the Jan Mayen area' Marine Policy 9 (1985) 16-38.

Churchill, R.R. and A.V. Lowe The Law of the Sea. (Manchester: Manchester University Press, 1983). (Second edition) (Manchester: Manchester University Press, 1985).

Clain, L.E. 'Gulf of Maine - a disappointing first in the delimitation of a single maritime boundary' Virginia Journal of International Law 25 (1985) 521-620.

Colson, D.A. 'The UK-France continental shelf arbitration' American Journal of International Law 72 (1978) 95-112.

Couper, A.D. (Ed.) The Times Atlas of the Oceans. (London: Times Books, 1983).

Crosby, D.G. 'The UNCLOS III definition of the continental shelf: application to the Canadian offshore' in D.M. Johnston and N.G. Letalik (Eds) The Law of the Sea and Ocean Industry: New Opportunities and Restraints 473-486. (Hawaii: Law of the Sea Institute, University of Hawaii, 1984).

Day, A.J. (Ed.) Border and Territorial Disputes. (Harlow; Keesing Reference publication, Longman,

Select Bibliography

1982).

Denman, D.R. Markets Under the Sea (London: Institute of Economic Affairs, 1984).

Dillon, W.P and D.G. Howell 'Nonliving EEZ resources; minerals, oil and gas' Oceanus 27 (1984/85) 31-32.

Douglas, R.P.A. Harbour Law. (2nd edition) (Lloyds of London Press Ltd, 1983).

Draft Environmental Impact Statement: Proposed Outer Continental Shelf Polymetallic Sulfide Minerals Lease Offering, Gorda Ridge Area Offshore Oregon and Northern California. (Reston, Virginia: Minerals Management Service, Department of the Interior, 1983).

Drysdale, A.D. and G.H. Blake The Middle East and North Africa: a Political Geography. (New York: Oxford University Press, 1985).

Dzurek, D.J. 'Boundary and resource disputes in the South China sea' in E.M. Borgese and N. Ginsburg (Eds) Ocean Yearbook 5 254-284. (Chicago: Chicago University Press, 1985).

Earney, F.C.F. Petroleum and Hard Minerals From The Sea. (London: Arnold, 1980).

El-Hakim, A.A. The Middle Eastern States and the Law of the Sea. (Manchester: Manchester University Press, 1979).

Feldman, M.B. 'The Tunisia/Libyan continental shelf case: geographic justice or judicial compromise?' American Journal of International Law 77 (1983) 219-238.

'Fisheries Case (United Kingdom v Norway), Judgment of December 18th, 1951' Report of Judgments, Advisory Opinions and Orders, 1951, 116-208. (The Hague: International Court of Justice, 1951).

Fisheries of the European Community. (Edinburgh: White Fish Authority, 1977).

Francioni, F. 'Legal aspects of mineral resource

Select Bibliography

exploitation in Antarctica'. Paper presented to the
regional meeting of the American Society of
International Law, Cornell University, 7-8 October
1985.

Frank, R.F. 'The convention on the conservation of
Antarctic marine living resources' Ocean Development
and International Law 13 (1983) 291-346.

Friedmann, W. 'The North sea continental shelf cases
- a critique' American Journal of International Law
61 (1970) 229-240.

Fulton, T.W. The Sovereignty of the Sea. (Edinburgh
and London; William Blackwood and Sons, 1911.
Reprinted in Millwood, New York: Kraus Reprint Co.,
1976).

Gamble, J.K. Jnr (Ed.) The Law of the Sea: Neglected
Issues. (The Hague: Law of the Sea Institute,
University of Hawaii, 1979).

Gardiner, P.R.R, R.P. Riddihough, and K.W. Robinson,
'The Law of the Sea, seabed resources and Ireland'
Technology Ireland 6 (1974) 1-6.

Gardiner, P.R.R. and K.W. Robinson, 'The Law of the
Sea: the continental shelf and hydrocarbon
resources' Technology Ireland 9 (1977) 7-12.

Gardiner, P.R.R. 'Reasons and methods for fixing the
outer limit of the legal continental shelf beyond
200 nautical miles' Iranian Review of International
Relations 11/12 (1978) 145-170.

Geographic Notes Nos 1-4, (Washington, DC: US
Department of State, Bureau of Intelligence and
Research, Office of the Geographer, 1985-86).

Gibson, J. 'Coastal zone management Law'. Paper
presented to the Greenwich Forum Conference: 'Perils
and prospects of the present inshore and offshore
regime for Britain's coasts and coastal zone. Is it
time for change?' Greenwich VIII, London 1962.

Gibson, J. 'Foreshore: a concept built on sand'
Journal of Planning and Environmental Law (1977)
762-770.

Gibson, J. 'Coastal zone management law: a case

study of the Severn Estuary and the Bristol Channel' Journal of Planning and Environmental law (1980) 153-165.

Glassner, M.I. 'The law of the Sea' Focus 28 (1978) 1-24.

Grisel, E. 'The lateral boundaries of the continental shelf and the judgment of the International Court of Justice in the North Sea Continental Shelf Cases' American Journal of International Law 64 (1970) 562-593.

Gulland, J. 'The new ocean regime: winners and losers' Ceres 12 (1979) 19-23.

Hannessian, J. Jnr. 'National interests in Antarctica' in T. Hatherton (Ed.) Antarctica. (London: 1965).

Hayashi, M. 'The Antarctic Question in the United Nations'. Paper presented to the Regional Meeting of the American Society of International Law, Cornell University, 7-8 October 1985.

Hatherton, T. (Ed.) Antarctica. (London: Methuen, 1965).

Hedberg, H.D. National-International Jurisdictional Boundary on the Ocean Floor. (Kingston, Rhode Island: Law of the Sea Institute, University of Rhode Island, Occasional Paper No. 16, 1972).

Hedberg, H.D. 'Ocean boundaries and petroleum resources' Science 191 (1976) 1009-1018.

Hedberg, H.D. 'Ocean floor boundaries' Science 204 (1979) 135-144.

Herman, L.L. 'The Court giveth and the Court taketh away: an analysis of the Tunisia - Libya continental shelf case' International and Comparative Law Quarterly 33 (1984) 825-858.

Heskin, A. 'A profile of the Irish fishing industry' in Fisheries of the European Community. (Edinburgh; White Fish Authority, 1977).

Hodgson, R.D. and L.M. Alexander Towards an Objective Analysis of Special Circumstances: Bays,

Rivers, Coastal Archipelagos and Atolls. (Kingston, Rhode Island: Law of the Sea Institute, University of Rhode Island, Occasional Paper No. 13, 1972).

Hodgson, R.D. and J.E. Cooper 'The technical delimitation of a modern equidistant boundary' Ocean Development and International Law 3 (1976) 361-388.

Hodgson, R.D. and R.W. Smith 'The informal single negotiating text (Committee II): a geographical perspective' Ocean Development and International Law 3 (1976) 225-259.

Hodgson, R.D. and R.W. Smith 'Boundary issues created by extended national marine jurisdiction' Geographical Review LXIX (1979) 423-433.

Hudson, M.D. 'The thirty-second year of the World Court' American Journal of International Law 48 (1954) 1-22.

Hutchison, D.N. 'The concept of natural prolongation in the jurisprudence concerning delimitation of continental shelf areas' British Yearbook of International Law 57 (1984) 133-187.

I.C.J. Pleadings, The Minquiers and Ecrehos Case (United Kingdom/France). 2 Vols. (Leyden: A.W. Sijthoff, 1955).

International Boundary Study, Series A, Limits in the Seas. Nos 1-105. (Washington, DC:US 'Department of State, Bureau of Intelligence and Research, Office of the Geographer 1970-1986).

Jogota, S.P. Maritime Boundary. (The Hague: Martinus Nijhoff, 1985).

Johnson. D.H.N. 'The Minquiers and Ecrehos Case' International and Comparative Law Quarterly 3 (1954) 189-216.

Johnston, D.M. and N.G. Letalik, (Eds.) The Law of the Sea and Ocean Industry: New Opportunities and Restraints. (Hawaii: Law of the Sea Institute, University of Hawaii, 1984).

Karmon, Y., A. Shmueli, and G. Horowitz, (Eds.) The

Selected Bibliography

Geography of the Mediterranean Basin. (Tel Aviv:
Ministry of Defence, 1983). (In Hebrew).

Kent, G. and M. Valencia (Eds.) Marine Policy in
Southeast Asia. (Los Angeles: University of
California Press, 1985).

Kirman, L.P. The White Road, A Survey of Polar
Explorations. (London: Hollis and Carter, 1959).

Kliot, N. 'The unity of semi-landlocked seas' in Y.
Karmon, A. Shmueli and G. Horowitz (Eds.) The
Geography of the Mediterranean Basin. (Tel Aviv:
Ministry of Defence, 1983). (In Hebrew).

Law of the Sea Bulletin Nos 1-7. (Office of the
Special Representative of the Secretary General for
the Law of the Sea, New York: United Nations
1985-86).

Legault, L.H. amd B. Hankey, 'From sea to seabed:
the single maritime boundary in the Gulf of Maine
case' American Journal of International Law 79
961-991.

Luciani, G. (Ed.) The Mediterranean Region: Economic
Interdependence and the Future of Society. (London:
Croom Helm, 1984).

Luciani, G. 'The Mediterranean and the energy
picture' in G. Luciani (Ed.) The Mediterranean
Region: Economic Interdependence and the Future of
Society 1-40. (London: Croom Helm, 1984).

McDorman, T.L., K.P. Beauchamp, and D.M. Johnston,
Maritime Boundary Delimitation, (Lexington:
Lexington Books, 1983).

McDorman, T.L., P.M. Saunders, and D.L. Vanderzwaag,
'The Gulf of Maine boundary: dropping anchor or
setting a course?' Marine Policy 9 (1985) 90-107.

McGregor, B.A. and M. Lockwood, Mapping and Research
in the Exclusive Economic Zone, (Reston, Virginia:
1985).

McKelvey, V.E. 'Interpretation of the UNCLOS III
definition of the continental shelf' in D.M.
Johnston, and N.G. Letalik, (Eds.) The Law of the

Select Bibliography

Sea and Ocean Industry: New Opportunities and Restraints 465-472, (Hawaii: Law of the Sea Institute, University of Hawaii, 1984).

McRae, D.M. 'Delimitation of the continental shelf between the UK and France: The Channel arbitration' Canadian Yearbook of International Law 15 (1977) 173-197.

Mangone, G.J. (Ed.) American Strategic Minerals, (New York: Crane Russak, 1984).

Marshall, H.R. 'Disputed areas influence OCS leasing policy' Offshore 45 99-100. (Tulsa: Penwell Publishing Co.).

M'Cormick, R. Voyages of Discovery. 2 Vols. (London: Sampson Law, 1884).

'Mediterranean states stifled by disputes' Offshore June 20, 1979, 159-165. (Tulsa: Penwell Publishing Co.).

Moore, J.R. 'Alternative sources of strategic minerals from the seabed' in G.J. Mangone, (Ed.) American Strategic Minerals 85-108. (New York: 1984).

Moore, J.R. 'OSIM for marine mining' Marine Mining 5 (1986) 335-336.

Mountfield, D. History of Polar Exploration. (London: Hamlyn, 1974).

O'Connell, D.P. The International Law of the Sea Vol.1. (Edited by I.A. Shearer) (Oxford: Clarendon Press, 1982).

O'Connell, D.P. The International Law of the Sea Vol.2. (Edited by I.A. Shearer) (Oxford: Clarendon Press, 1984).

Ostreng, W. 'Regional delimitation agreements in the Arctic Seas: cases of procedure?'. Paper presented at the 18th Annual Conference of the Law of the Sea Institute, San Francisco, USA, 1984.

Peck, D.L. 'The U.S. Geological Survey program and plans in the EEZ' in Symposium Proceedings: A

National Program for the Assessment and Development of the Mineral Resources of the United States Exclusive Zone, Nov. 15, 16, 17, 1983. (Alexandria, Virginia: USGS Circular 929, 1984).

Pendley, W.P. 'Importance of the EEZ Proclamation' in Symposium Proceedings: A National Program for the Assessment and Development of the Mineral Resources of the United States Exclusive Economic Zone, Nov. 15, 16, 17, 1983. (Alexandria, Virginia,: USGS Circular 929, 1984).

Pharand, D. The Law of the Sea of the Arctic. (Ottawa: 1973).

Pharand, D. 'The implications of changes in the law of the sea for the "North American" Arctic Ocean' in J.K. Gamble, Jnr. (Ed.) The Law of the Sea: Neglected Issues, (The Hague: Law of the Sea Institute, University of Hawaii, 1979).

Pierce, G.A.B. 'Selective adoption of the new law of the sea: the United States proclaims its exclusive economic zone' Virginia Journal of International Law 23 (1984) 586-601.

Prescott, J.R.V. The Political Geography of the Oceans. (New York: John Wiley and Sons; Newton Abbot, London and Vancouver; David and Charles, 1975).

Prescott, J.R.V. 'Existing and potential maritime claims in the Southwest Pacific Ocean' in E.M. Borgese, and N. Ginsburg, (Eds.) Ocean Yearbook 2 317-345. (Chicago: Chicago University Press, 1980).

Prescott, J.R.V. 'Boundaries in Antarctica' in Australia's Antartic Policy Options, (Canberra: Centre for Resource and Environmental Studies, Australian National University, 1984).

Prescott, J.R.V. 'Maritime jurisdictional issues' in G. Kent, and M. Valencia, (Eds.) Maritime Policy in Southeast Asia 58-97. (Los Angeles: University of California Press, 1985).

Proposals for a U.K. Fishery Policy. (Hull: British United Trawlers Limited, 1976).

Quigg, P.M. A Pole Apart: The Emerging Issue of

Select Bibliography

Antarctica. (New York: McGraw Hill, 1983).

Regnier, J. 'The real meaning of Community' in _Fisheries of the European Community_. (Edinburgh: White Fish Authority, 1977).

Rhee, S-M. 'Equitable solutions to the maritime boundary dispute between the United States and Canada in the Gulf of Maine' _American Journal of International Law_ 75 (1981) 590-628.

Rhee, S-M. 'Sea-boundary delimitation between states before World War II' _American Journal of International Law_ 76 (1982) 555-588.

Rich, R. 'A minerals regime for Antarctica' _International and Comparative Law Quarterly_ 31 (1982) 717-718.

Schneider, J. 'The Gulf of Maine Case: the nature of an equitable result' _American Journal of International Law_ 79 (1985) 539-577.

Shaw, D.F. 'Conservation and development of marine and coastal resources' in _The Conservation and Development Programme for the U.K._. (London: Kogan Page, 1983).

Shusterich, K.M. 'Arctic issues coming to fore' _Oceanus_ 27 (1984/85) 82.

Slade, D.C. 'Maritime boundaries of the United States' in _The Exclusive Economic Zone of the United States: Some Immediate Policy Issues_. (Washington, DC: National Advisory Committee on Oceans and Atmosphere, 1984).

Smith, H.D. and C.S. Lalwani _The North Sea: Sea Use Management and Planning_. (Cardiff: Centre for Marine Law and Policy, UWIST, 1984).

Smith, R.W. 'The maritime boundaries of the United States' _Geographical Review_ 81 (1981) 395-410.

Smith, R.W. 'A geographical primer to maritime boundary-making' _Ocean Development and International Law_ 12 (1982) 1-22.

Smith, R.W. (Ed.) 'National Claims to Maritime

Select Bibliography

Jurisdiction' Limits in the Seas No. 36 - 5th Revision. (United States Department of State Bureau of Intelligence and Research, Office of the Geographer, 1985).

Smith, R.W. Exclusive Economic Zone Claims: An Analysis and Primary Documents. (Dordrecht, The Netherlands: Martinus Nijhoff Publishers, 1986).

Sohn, L.B. and K. Gustafson, The Law of the Sea in a Nutshell. (St Paul: West Publishing Co., 1984).

Symmons, C.R. 'The Rockall dispute' Irish Geography 8 (1975) 122-126.

Symmons, C.R. The Maritime Zones of Islands in International Law. (The Hague: Martinus Nijhoff Publishers, 1979).

Symmons, C.R. 'The outstanding maritime boundary problems between Ireland and the U.K.' Paper presented to the 19th Annual Conference of the Law of the Sea Institute, Cardiff, 1985.

Symmons, C.R. 'The Rockall dispute deepens: an analysis of recent Danish and Icelandic actions' International and Comparative Law Quarterly 35 (1986) 344-373.

Symposium Proceedings: A National Program for the Assessment and Development of the Mineral Resources of the United States Exclusive Economic Zone, Nov. 15, 16, 17, 1983. (Alexandria, Virginia,: USGS Circular 929, 1984).

The Durham Coast. (Durham County Council/Easington District Council, 1982).

The European Community's Fishery Policy. (Luxembourg: Commission of the European Communities, 1985).

The Exclusive Economic Zone of the United States: Some Immediate Policy Issues. (Washington, DC: National Advisory Committee on Oceans and Atmosphere, 1984).

'The Minquiers and Ecrehos Case, Judgement of November 17th, 1953' Reports of Judgments, Advisory Opinions and Orders, 1953 47-111. (Leyden: A.W.

Sijthoff, 1954).

'The North Sea Continental Shelf Cases' Reports of Judgments, Advisory Opinions and Orders, 1969. (The Hague: International Court of Justice, 1969) Reprinted in International Legal Materials VIII (1969) 340-433.

Traavik, K. and W. Ostreng, 'Security and ocean law: Norway and the Soviet Union in the Barents Sea' Ocean Development and International Law 4 (1977) 343-367.

'Tribunal Arbitral Pour la Délimitation de la Frontière Maritime Guinée/Guinée-Bissau, 14 Février 1985' Révue Générale de Droit International Public 89: (1985) 484-537. English translation in International Legal Matérials XXV (1986) 251-307.

Troy, K. 'The making of offshore boundaries: beyond the Gulf of Maine - Part 1' Oil and Gas Law and Taxation Review 11 (1984/85) 289-298.

Troy, K. 'The making of offshore boundaries: beyond the Gulf of Maine - Part II' Oil and Gas Law and Taxation Review 12 (1985) 314-328.

Underal, A. The Politics of International Fisheries Management: the Case of the Northeast Atlantic. (Oslo: Universitetsforlaget, 1980).

United Nations The Law of the Sea: Official Text of the United Nations Convention on the Law of the Sea. (London and Canberra: Croom Helm: New York: St Martin's Press, 1983).

Westermeyer, W.E. The Politics of Marine Resource Development in Antarctica. (Boulder, Colorado: Westview Press, 1984).

Wise, M. The Common Fisheries Policy of the European Community. (London and New York: Methuen, 1984).

Zorn, S.A. 'Antarctic minerals: a common heritage approach' Resources Policy 10 (1984) 2-18.

adjacent   coastlines/states   5,16,18,20-22,27,45,46,
58,60,152,155,168,190
Adriatic Sea 214
Aegean Sea 8,210,211,216-219,223
Africa 5,79
    North 78,208
    south of the Sahara 78
    West 32,33
Ago, Judge Roberto 183
Alaska 169, 170, 174
Alaska Boundary Tribunal Award 1903 169
Albania 210-212, 217, 218, 221-223
    and Greece 216
    and Italy 216
    straight baselines 41
Alexander, L.M. 40,108
Algeria 211, 217,221,222
Americas 5
Anguilla (UK) 168,173
Antarctic maritime boundaries 227-240
Antarctic Treaty 1959 233-240
Antarctic Ocean 155
Antarctica 1,61,78,227-240
    living resources 233,235-237
    mineral resources 237-240
Antigua 47
    and Barbuda 166
Arab states 210
archipelagic baselines 5,38-49,105
archipelagic boundaries 5
archipelagic states 5,45-48,64
archipelagos 45-48,105,156
Arctic Ocean 7,11,39,78,154,155,231
Area, the see international seabed
Argentina 165,231,234,237,240
    and Gulf of Rio de la Plata 39

Atlantic Ocean 11,183,213,218,221,222
    North 78,82,84,86; Northeast 83,84,125,135,139,
    144; Northeast/Arctic 7, Northwest 7; Southwest
    7,; Southeast 7;
Australia 26,41,45,47,78,229-231,234,236,237,240
Australian Antarctic Territory 231,232
Azores 83,122,130

Bahamas 47,168,172,173
Bahrain 60
    and Saudi Arabia joint zone 9,10
Baker Island (USA) 174,175
Baltic Sea 7,11,78
Bangladesh 166
    straight baseline 45,65
Barbados 166
Barents Sea 147-159
    continental shelf boundary 147-157
    exclusive economic zone boundary 147,151-156
    fisheries 148,151,157 158
    Grey Zone Agreement 1978 151-152,157,159
Basdevant, Judge 97
baselines 23,53-58,59,61,64,70,78,93,163,209,239
    archipelagic, see archipelagic baselines
    closing lines, see closing lines
    normal 38
    straight, see straight baselines
basepoints 41,44,45,47,107,108
Bass Straight 45
Bay of Biscay 128,129
Bay of El Arab 212
Bay of Fundy 188
Bear Island 83,147,150,156
Beaufort Sea 169,170,174
Belgium 83,127,234
Belize 7
Bering Sea 169,171,174
Black Sea 7,11,78,211,218
Blake, G.H. 7,216,218
boundaries:
    single   (line)   15,26,28,31,182   see   also
    continental shelf and EEZ
    types of boundary 5,77
boundary delimitation agreements 1,2,3-4,6,8-11
    Channel Islands and France 89-111
    in the Caribbean 6-7
Bowett, D.W. 104
Brazil 78,234,237
Briggs, Professor 22
British Columbia 169
Brown E.D. 69

Brunei 1
Burma 41,45,166

California 174-176
Canada 5,26,28,30,39,78,178,183 see also cases:
Canada-United States
    boundaries with US 168,169,170,183
    fisheries 27,183
Canary Islands 83
Cape Verde 47,166
Caribbean 6,7,11
    boundaries in Eastern Caribbean 6
    boundaries with USA 172-173
    fisheries 7
    hydrocarbon and mineral resources 7
cases:
    Canada-United States 1984 (Gulf of Maine
    Maritime boundary) 5,9,24-29,31,153,155,169,
    182-204
    Guinea-Guinea Bissau 1983 (Maritime boundary)
    9,29,31-33,154
    Libya-Tunisia 1982 (Continental Shelf) 8,9,
    22-25,58,61,62,214
    Libya-Malta 1985 (Continental Shelf) 8,9,29-31,
    53,153,154,215
    North Sea 1969 (Continental Shelf) 8,16-20,22,
    24,26,78
    Norway-United   Kingdom   1951   (Fisheries
    jurisdiction) 40
    United Kingdom-France 1977 (Continental Shelf)
    16,18-22,26,105-108,219
Celtic Sea 134,140,141,143
Champ, M.A. 178
Channel Islands 19,21 see-also cases: United
Kingdom-France; and conventions: Anglo-French
Fisheries and London European fisheries
    continental shelf 21
    exclusive fisheries' limits/rights/zones 91-
    100, 104-106
    fishing industries 90,110
    mer commune (common fisheries) 90-96,98-108,110
    Minquiers and Ecrehos 92,94-98,99,104,105,
    107,108
    oyster fishing 90,91,93,95,96,110,111
    special regime area with France 91,93,96,99,
    100,107
    territorial sea 91,93-98,104-108
    1928 UK-France Fisheries Agreement 96,99,111
    1951 UK-France Fisheries Agreement 92,95-99,
    104,107,110,111

1968 UK-France Fisheries Agreement (draft) 100,101
Chile 39,165,231,234,237,240
    straight baseline 44
China 12,78,79,234
Chukchi Sea 169,170,174
closing lines for rivers and bays 38,39,41,53,61, 212,254
    Gulf of Maine 28,190,191,194,195
    Gulf of Sirte 39,49,61,213,216
Cohen, Judge ad hoc Maxwell 183
Columbia 7,41,65,166
    Columbian-Dominican Republic joint zone 10
    straight baseline in Pacific Ocean 42,43
common zone, see zones of joint economic exploitation
Comoros 47,166
contiguous zones 78,163,208,209,220-222
continental shelf 19,25,27,29-31,77,78,107,134-136, 140,143,164,185,187,188,195,208,211,213-215,220,239, 251,
    boundaries 2,8,31,106
    Commission on the limits of 68-71, text 74-76, 144
    definitions 23, text of UNCLOS Article 76:72-74
    200 mile limit 29,31,136
    margin(s) 1,48,49,65-68,70,137-138,149,174,175
    outer limit 64,65-71
    single boundary with EEZ 15,26,28,153, 156, 158
    conventions:
    1839 Anglo-French Fisheries Convention 90-94, 96,99,103,110,111
    1867 Anglo-French Fisheries Convention 93,94
    1867 United States-Russia Convention 169-171
    1886 France Portugal (Guinea-Guinea Bissau) 31,45
    1958 Geneva Continental Shelf Convention 15-16, 18-20, 26,133,148,149,153,156,157,167,212,214
    1958 Convention on the Territorial Sea and Contiguous Zone 38,40,49,173
    1964 London European Fisheries Convention 98,105,111
    1982 UN Convention on the Law of the Sea 2,12,15,30,31,34,38,39,49,52,63-65,68-70,72-76, 84,87,88,105,136,144,149,164,166-168,209-212, 214,220,222,223,238,239
        Annex II Commission on the limits of the Continental Shelf 68-71, (text 74-76),144
        Article 7:40-46,48,49
        Article 47: 46,48,64

Article 74: 2,30
Article 76 Definition of the Continental
Shelf: 1,30,65-67,70,71, (text 72-74),138,
139,140
Article 83: 2,30
Cook Islands 166,168,174
Corsica 217
Costa Rica 166
Courts of Arbitration
    Chile-Argentina 44
    UK-France 1977 16,18-22,105-106
Cuba 7,41,47,78,166,168,172,173
Cyprus 211,216,217,220-222
    Britain's bases 8,208,218
    occupied Northern 8
    and Turkey 216

Denmark 16-18,44,83,118
    territorial claim 138-140
Demarffy, A. 166
Democratic Kampuchea 166
Democratic Yemen 166
Dillon W.P. 178
Djibouti 166
Dominica 166
Dominican Republic 166,168,173
    Columbia-Dominican Republic joint zone 10
Drysdale, A.D. 7,216

East Asian Seas 7,11
Ecuador 43,165
    Galapagos Islands 45,165
Egypt 211,217,220-222
    and the Bay of El Arab 212
    straight baselines 212
English Channel 20,129,144
Equatorial Guinea 166
equidistance or median line 16,19,21,23,30,169,170,
172,185,210,214
equidistance principle 15-20,26,32,33,156,169,172,
173,191,214,215,218
equitable criteria/principle 16-18,20-24,26,27,29-
31,33,60,153,187-188,214,218
equitable solution/result 2,15,17,18,20,23-30,32,33,
52,144,153,155,188,190-192,194,196,201,203,213
Europe 78,79,87
European Economic Community 87,103,117-130.144,223
    Common Fisheries Policy 103,117-121,125-129,143
    European Commission 117,123,124
    Fishing limits 117-130

    national fishery zones 118-127
    200 mile limit 121-124
    Treaty of Accession 1972 120,127
    Treaty of Rome 1957 118,119,125
Exclusive   Economic   Zone   2,12,19,25,29,30,39,41,
48,77-79,81,87,121,122,124,144,163,165,167,168,209,
213,220-223,236,238,239
    boundaries 31
    legislation by UNCLOS III 12,31,41,49,64,167
    outer limit 64-65,68,69-71,83,167
    single boundary with Continental Shelf 15,26,28

Faroe Islands 83,119,122,135,138,139,142   see also
Rockall-Faroe Plateau
Federated States of Micronesia 168,174
Fiji 47,166
Finland 150
    straight baseline 44
fisheries 78,84,86,245 see also particular regions
    on the Georges Bank 27,169,185,191,194,202
    management   80,81-82,85,87,88,117,123,125,126,
    167,194,250
    UK-USA Reciprocal Fisheries Agreement (Virgin
    Islands) 173
fisheries zones 166,221-223
    boundaries 2
    of Canada 183
    of the Channel Islands 21,89-111
    in the European Community 117-130
    Exclusive Fisheries Zones 29,30,78,81,194,222,
    223
    limits 77,82
    of the USA 164,165,173,183
Florida 172
France   8,19,83,92,127,128,144,162,166,183,211,213,
217,219,220,221,222,231,234,236,237 see also cases:
United Kingdom-France, and Channel Islands
    and Italy 216,218
    and Spain 214,216,218
    and baselines 43,212
    territorial seas 91,93,104-107

Gaza strip 8
Georges Bank 182-204
Gibraltar 8,208,217,219-221
Goud, M.R. 166
Gordon, W.G.170
Greece 61,210,211,213,217,221,222
    and Albania 216
    and Italy 214,216,218
    and Libya 216

and Turkey 208,216,218-219
Greenland 5,83,119,123,139,147,227
Grenada 47,166
Gros, Judge Andre 28,183,192,194
Guatemala 7,166
Guinea 32,43,166
Guinea Bissau 32,166
    straight baseline 45
Gulfs:
    Golfo   de   Cupica,   see   Columbian   straight
    baseline
    Gulf of Bothnia 44
    Gulf of California 176
    Gulf of Gabes 213
    Gulf of Maine 27,183,184,187-189,203
    Gulf of Mexico 6,7,172,174
    Gulf of Oman 44
    Gulf of Sirte (Khalij Surt) 8,39,61 see also
    closing lines
    Gulf of Taranto 213
    Gulf of Thailand 12
    Persian Gulf, see Persian-Arabian Gulf
    at mouth of Rio de la Plata 39
Gulf Stream 199-200
Gutting, R.E. 170

Haiti 46,47,166
Hawaiian Islands 164,175,177
high seas 32,41,49,98,106,107,194,236,237
historic bay claims 8,61,212,213
Hodgson R.D. 40
Honduras 165,166
Hong Kong 3-4
Howel, D.G. 178
Howland Island (USA) 164,174,175
hydrocarbon resources (oil and gas) 6,8,78,85,110,
133,134,136,140,141,143,148,151,157,169-171,175,215,2
19,223,245,251-252

Iceland 47,68,83,123,135,142,165,166
    Iceland-Norway joint zone 3-4,9,10
    offshore territorial claim 138-140
    straight baselines 41,43-44,65
India 45,78,166,234
Indian Ocean 7,11
Indonesia 47,48,166
internal waters 39,42,48,49,77,208,213,220,245,246
    regime of 41,105
International Court of Justice 5,15,22,23,26,34,40,
41,52,62,95-98,107,149,169,200,203,213-215,218

Gulf of Maine Chamber 29,182,183,185,186,188-195
international law 11,16,17,19,22,24-29, 32, 33,63, 140, 141,152-157,166,177,183,186,191,231,236,238,240
    Statute of the International Court of Justice 2,23,213
International Law Commission (1956 report) 40
international seabed (the Area) 1,49,63-76,84,87,239
International Seabed Authority 68,144,239
Ionian Sea 60,217
Iran
    and Saudi Arabia 10
    straight baselines 43,44
    and United Arab Emirates 12
Iraq 45
Irish Sea 103,140,143
islands 60,61,68,144,167,172,208,210,214,236 see also archipelagos
    causing delimitation problems 218,219 see also Channel Islands
    disputes over 2,6,11-12
    fringing a coast 40,43,45,212
    in the Guinea-Guinea Bissau case 32
    in the Gulf of Maine 28,187
    as special or relevant circumstances 24,33,156
    states 87
    used to draw straight baselines 43,48
Israel 208,210,211,217,220,221
Italy 8,25,29,61,183,211,213,214,217,220-223
    and Albania 216
    and France 216,218
    and Greece 214,216,218
    and Libya 216
    and Malta 30,215,216,219
    and Spain 214,216,218
    straight baselines 41,43,212-213
    and Tunisia 214,216,218,222
    and Yugoslavia 214,216,218,222
Ivory Coast 166

Jagota, S.P. 9,11,216
Jamaica 47
Jan Mayen Island 9,83,122
Japan 47,78,79,169,234,237
    Iwo Jima, boundary with USA 174
    Japan-Korea joint zone 3-4,9,10
    and USSR 12
Jarvis Island (USA) 164,174,175
Johnson D.H.N. 95
joint development zone, see zones of joint economic exploitation

Kenya 166
Kiribati 47,48,168,174
Kirk, William 192,193

Latin America 78,79
Lebanon 211,217,220-222
Libya 29,61,153,211,215,217,221-223 _see also_ cases:
Libya-Tunisia, Libya-Malta
    and Greece 216
    Gulf of Sirte 8,39,213,218
    and Italy 216
    and Malta 30,53-58,60,215,216
    and Tunisia 27,58-60,216
    and United States 49
low-tide elevations 41,47,61
low-water line/mark 38-40,44,91,93,173,212,244,245,
250,251,253,254
Luciani, G. 216

Madagascar 47,166
Madeira 83
Malaysia 12,46,240
Maldives 47
Malta   25,29,30,38,47,60,61,211,220-223   _see   also_
cases: Libya-Malta
    baselines 38,43,53-57,213
    island of Filfla 38,55,56,213
    island of Gozo 54,55,57,58
    and Italy 215,216,219
    and Libya 30,53-58,215,216
    and Tunisia 215
management,
    environmental 84,87,88
    of fisheries, _see_ fisheries
    of mineral resources 82-84,87
    of shipping and navigation 80-81,87
    of strategic affairs 80,81,82,87
Marshall Islands 168,174
Mariana Islands 164,174,175
Mauritania 41,83,166
Mauritius 47,166
McGregor, B.A. 166
median line(s) 2,21,30,44,60,83,107-109,121,122,124,
139,150,153, 155,156,190,191,214,219 _see also_ equi-
distance line
    hypothetical 3-4,6
Mediterranean Sea 5,7,30,60
    boundaries 8,11,208-223
Mexico 166,177
    boundaries with USA 168,172
mineral and aggregate resources 80,85,87,98,175-

177,245,251,253
Monaco 8,211,217,221,222
Morocco 1,83,166,211,217,220-222
    and Spain 208,216,219,220
Mosler, Judge Hermann 183
Mozambique 166

Namibia 1
NATO 148,219
natural boundaries 62,196,199
natural prolongation 17,18,20-25,29,31,33,58,66,68,
107,136,137,141,210
near-shore zone 253
Netherlands 16-18,83,171,173
New Caledonia 48
New Zealand 47,78,79,166,174,229,230,234,237,238,240
    Tokelau 168,174
Nicaragua 185
Nigeria 166
Niue 166
North Korea 46,79
North Sea 11,124,129,133,136,157,171
Norway 9,10,39-41,44,83,118-121,124,125,130,166,230-
232,234,236
    fishing limits 119-120
    Iceland-Norway joint zone 3-4,9,10
    Norwegian-Soviet Fishery Commission 151
    straight baseline 40
    and the USSR in the Barents Sea 147-159

ocean management 72-88
offshore zone 235,255
Oman 166
opposite coastlines/states 5,16,18,20-22,27,52-58,
152,156,168,190
Oregon 174,176

Pacific Islands 78,164,174-176
Pacific Ocean 5,11,42,43,172
    Eastern 7; Western and Central 7; North 84,169;
    Northwest 78
Pakistan 45,166
Panama 42,43
Papua New Guinea 26,47
Peck, D.L. 174
Persian-Arabian Gulf 5,6,7,11,44
Peru 165
Philippines 12,47,48,166
Poland 234
Portugal 31,83,104,128,166 see also conventions:
France-Portugal

Index

Prescott, J.R.V. 5,7,12,65,210,211,213,215,216,221
proportionality 17,20,22,25,33,34,52,153,155

Red Sea 7,9,11,158,210
relevant circumstances 24-27,30,31,33,153,194
Republic of Ireland (Eire) 19,22,41,47,83,103
    Continental Shelf Act 1968 133
    exclusive fishing zone 122,129,141,142
    fisheries in the European Community 118,119,
    121-127,129
    offshore jurisdictional claims 133-144
Republic of Korea 166
    Japan-Korea joint zone 3-4,9,10
Rockall 83,134-136,138-143
Rockall-Faroe Plateau 134-138,140
Rowland, R.W. 166,170

St Vincent and the Grenadines 47
Samoa (American) 164,174,175
Samoa (Western) 47,166,168,174
Sao Tome and Principe 47,166
Saudi Arabia
    and Bahrain joint zone 9,10
    and Iran 10
    Saudi Arabian Sudanese Red Sea 'common zone'
    3-4,9,10,158
Schwebel, Judge Stephen 183
Scilly Isles 21-22
Scotland 129,141,252
sector line 149-151,154-156
Senegal 32,41,166
Seychelles 47,166
Shetland/Orkney restricted zone 122,127
Sierra Leone 32
Singapore 1,47
Smith, R.W. 5,7,9,11,221
Solomon Islands 47,166
South Africa 78,234,236
South America 229-231
South China Sea 12
    Spratley and Paracel Islands 12
Spain 83,104,127-129,166,211,217,219-222
    and France 214,216,218
    and Italy 214,216,218
    and Morocco 208,216,219
    North Africa possessions 8: Ceuta and Melilla
    208,217,219,220; Islas Chaferinas 219
    straight baselines 213
    and the UK 208
special circumstances 2,15,16,19-22,26,153,156, 201,
214,219

Sprout, Harold and Margaret 193
Sri Lanka 47,166
straight baselines 38-49,53,65,104,105,172,173,211-213 _see also_ archipelagic baselines and closing lines
straits (international) 209
    Strait of Malacca 12
    Torres Strait 28
Sudan, Saudi Arabian Sudanese Red Sea 'common zone' 3-4,9,10,158
Suriname 166
Sweden 41,44,83,150
Syria 210,211,217,218,220-223
    straight baselines 212
    and Turkey 216

Taiwan 47
territorial waters/seas 32,39,42,43,47-49,64,65,77,81,103,133,163,185,208,210,211,214,220,244-247,250,254
    boundaries 2,9,31,46,152
    limits 12,209,254
    outer limit 41,46
Togo 166
Tonga 47,166,168,174
Trinidad and Tobago 47
Truman Proclamation of the United States 1945 16\
Tunisia 211,215-217,219,221-223 _see also_ cases: Libya-Tunisia
    and the Gulf of Gabes 213
    historic fishing rights 24
    and Italy 214,216,218,222
    and Libya 27,58-60,216
    and Malta 215
    straight baselines 213
Turkey 210,211,213,217,221
    and Cyprus 216,220
    and Greece 8,208,216,218-219
    and Syria 216
    straight baselines 213
Tuvalu 47
Tyrrhenian Sea 217,218

Union of Soviet Socialist Republic 78,166,170,213,233,234,238,240
    EEZ overlaps with USA 169-171
    and Japan 12
    and Norway in the Barents Sea 147-159
    Norwegian-Soviet Fishery Commission 151
United Arab Emirates 1,6
    and Iran 12

United Kingdom 19,47,81,83,118,119,121,123-128,135,
136,138-140,171,175,208,213,216,219 see also cases:
Norway-United Kingdom, United Kingdom-France
    claims to Antarctica 229-231,234,236,237
    coastal zone of England and Wales 244-255
    and Cyprus 216,218
    fisheries limits 141-143
    and the Republic of Ireland 133,134,140-144
    and Spain 208,219,220
United Nations 80,86,233,240
    Food and Agriculture Organisation 82,86
    General Assembly 228,240
United Nations Conference on the Law of the Sea
(Third) 1973-1982 8,18,22,23,28,77,80-82,87,133,136-
138,163,165,209,210,236
United States of America 26-28,30,78,134,148,166,
183,213,230,233,234,238,240 see also cases: Canada-
United States
    boundaries with Canada 168,169,183
    boundaries in the Caribbean 7,172-173
    boundaries with Mexico 168,172
    boundaries in the Pacific 174
    boundaries with USSR 169-171
    Exclusive Economic Zone 162-179
    fisheries 26,169
    Fisheries Conservation and Management Act 1976
    164,168,169,172
    and Libya 49
    ocean mineral resources 162-165,174-177
    Proclamation 5030 162,166-168
Uruguay 39,234

Vanatu 47,48,166
Venezuela 7,166
    boundary with USA 173
Vietnam 12,41,43,166
Virgin Islands (UK) 168,173
Virgin Islands (USA) 173

West Germany 16,18,44,83,127,183,230,231,234,237,
238
Whittlesey, D.W. 111

Yukon Territory 169
Yugoslavia 211,217,221,222
    and Italy 214,216,218,222
    straight baselines 213
zones of joint economic exploitation 3-4,9,10